Automobile Sus

Automobile Suspensions

COLIN CAMPBELL
M.Sc., C.Eng., M.I.Mech.Engrs

LONDON
CHAPMAN AND HALL

First published 1981
by Chapman and Hall Ltd
11 New Fetter Lane, London EC4P 4EE

© 1981 C. Campbell

Printed in the United States of America

ISBN 0 412 16420 5 (cased edition)
ISBN 0 412 15820 5 (paperback edition)

British Library Cataloguing in Publication Data

Campbell, Colin

Automobile suspensions

1. Automobiles — Springs and suspension
I. Title
629.2'43   TL 257   80-41445
ISBN 412-16420-5
ISBN 412-15820-5 Pbk

TO THE MEMORY OF
MAURICE OLLEY
who introduced us to the benefits
of independent suspension

02311

Contents

We have had enough of action and of motion we,
Roll'd to starboard, roll'd to larboard,
When the surge was seething free.

Alfred Lord Tennyson

Preface

This book is an introduction to the elementary technology of automobile suspensions. Inevitably steering geometry must be included in the text, since the dynamic steering behaviour, road-holding and cornering behaviour are all influenced by the suspension design. Steering mechanisms and steering components are not covered in this book. This is not a mathematical treatise, but only a fool or a genius would attempt to design a motor vehicle without mathematics. The mathematics used in this book should present no problem to a first-year university student. SI units have been used in general, but for the benefit of those not familiar with them we have included in brackets, in many cases, the equivalent values in Imperial units. Many engineers regard the Pascal as an impractical unit of pressure. The author has therefore expressed pressures in bars (1 bar = 10^5Pa). A deviation from SI units is the use of degrees and minutes, instead of radians, to express camber, castor, roll angles, etc. This is still common practice in the motor industry.

No attempt has been made to make any stress calculations on suspension components. The automobile engineering student will have access to other textbooks on such subjects as strength of materials and theory of structures.

The author is grateful for technical information, photographs and drawings supplied by many car and component manufacturers. The following companies have been particularly helpful: Automotive Products Ltd, Citroën Cars Ltd, Datsun (UK) Ltd, Dunlop Ltd, Lotus Cars Ltd, Lucas-Girling Ltd, Magnesium Elektron Ltd (MELMAG wheels), Mercedes-Benz (UK) Ltd, Moulton Developments Ltd (Hydragas suspension), Porsche Cars (GB) Ltd, Tech Del Ltd (Minilite wheels).

Special thanks are due to the editor of *Motor* for his permission to publish data from 'The 150 m.p.h. Corner', a report (3 March 1979) in that excellent journal, giving measurements made with the assistance of the Cranfield Institute on the Lotus 79 when cornering at 2.05g radial acceleration.

C.C.

1
Wheels and Tyres

1.1 Suspension

The term *suspension* suggests 'supporting from above', as in 'suspension bridge' or 'suspender', since the earliest forms of suspension used on the roads did literally suspend the body of the horse-drawn carriage at the four corners by very flexible leaf springs with outer ends that curved over and upwards to join the shackle pins in the carriage frame.

In the modern automobile suspension plays a dual role. It 'suspends' the body on a shock-absorbing system and at the same time maintains the contact patches of the four tyres in effective contact with the road surface. One cannot overemphasize the importance of this second role. If we fail in this, the driver cannot control the vehicle. Rally drivers are well aware that effective road contact is sometimes lost. They wear safety harnesses and crash helmets to reduce the dangers from this eventuality and the backup crew carry a host of spares to repair the damage. Even so, a good suspension system under normal conditions keeps all four wheels in contact with the road surface and makes it possible for a driver of average ability to keep the vehicle under satisfactory directional control. A bad suspension system fails to do this under identical road conditions.

It is perfectly logical therefore to begin our analysis of suspension systems from the bottom, since the car's four areas of contact with the road surface made by the tyres (*footprints*, in American terminology) are so vital to the behaviour of the vehicle when in motion. This leads us then to the wheels and the tyres.

1.2 The wheel

Nobody knows who invented the wheel. That delightful American strip-cartoon series *BC*, suggested that their most

advanced-thinking   cave-dweller   announced   a   major
breakthrough when he dropped the idea of a square wheel in
favour of a triangular one. After all, such a wheel would only
give three bumps per revolution instead of four! Somebody
must have progressed towards a crude circle and from a
heavy, clumsy, solid wooden disk, held in place by pegs
driven into holes in the axletree, the wheel had developed
around the year 1800 B.C. into a spoked wheel. The early
Egyptian chariots used only four spokes, but the same
construction principle still survives in our few remaining
wheelwright's shops.

These craftsmen still make a central hub with radial spokes
surrounded by a felloe, or outer rim, made from several
interlocking arcuate pieces. One improvement on the
Egyptian design is the addition of a wrought-iron rim which
is heated in the forge and shrunk in place around the felloe.
These traditional carriage wheels and the wire-spoked steel
wheels that had been developed for the bicycle were the
alternatives available to the early automobile makers.

1.3   Modern automobile wheels

1.3.1   Steel wheels

More than 90% of modern wheels are of pressed steel. They
are strong and cheap to manufacture. They require negligible
maintenance and are only inferior to alloy wheels on one
count; they are heavier. Many alloy wheels are bought simply
to impress the neighbours. Even so, they are essential to the
serious rally competitor. All single-seater racing cars use
magnesium alloy wheels. The weights, or more correctly, the
mass of the unsprung components of a vehicle have a
significant influence on the performance of the suspension
system. As will be shown in Chapter 2 the suspension
behaviour is influenced favourably by a reduction in the
unsprung mass. The wheel, of course, is a substantial part of
this mass.

Steel wheels are made from two pressings, as shown in
cross section in Fig. 1.1. The inset distance and the rim
profile can be varied to suit the car manufacturer's

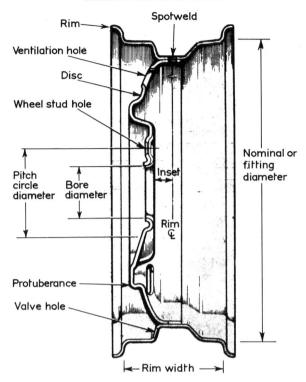

Fig. 1.1 Cross section of a steel wheel.

requirements. The flange profile, indicated by letters K, JK, J or C in the specification, is designed to comply with the tyre bead profile.

1.3.2 Aluminium alloy wheels

Aluminium alloy wheels are usually made from heat-treated castings. A typical example, the Minilite Sports wheel made by Tech Del Ltd of London, is shown in Fig. 1.2. The weight-saving compared with a steel wheel varies from about 30–50% the saving being greater for the wider wheels. The Minilite wheel is stove-enamelled to improve appearance and to provide corrosion protection.

Fig. 1.2 Minilite aluminium alloy wheel.

1.3.3 Magnesium alloy wheels

A typical aluminium alloy has a density of about one-third that of steel. Cast magnesium alloys are even lighter, having a density of slightly less than one-quarter that of steel. The tensile strength is 40–50% of a typical steel pressing, giving a great improvement in the *specific strength* (tensile strength per unit density). A cast magnesium alloy wheel of 5.5-in rim width and 13 in nominal diameter, as used on a typical works rally car weighs 4.2 kg in Minilite form. The weight of a steel wheel of this size can vary between 6.0 and 6.7 kg. Larger wheels yield a greater weight-saving. A 6JK × 15 steel wheel for a Jaguar weights 10.3 kg. A Minilite magnesium replacement weighs only 6.7 kg.

Rally wheels are designed to withstand very high shock loads. Since tyres are often ripped open by jagged rocks, the wheel rims must also suffer the same brutal treatment. Racing drivers do occasionally clip a kerb, particularly on circuits like Monaco. In general, however, wheels on racing cars do not suffer such high impact loads as those fitted to rally cars. Whereas a rim thickness of 6.4 mm will be used on a rally magnesium wheel, a single-seater racing car wheel is

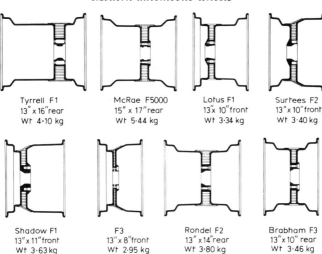

Tyrrell F1	McRae F5000	Lotus F1	Surtees F2
13″ x 16″rear	15″ x 17″rear	13″x 10″front	13″x 10″front
Wt 4·10 kg	Wt 5·44 kg	Wt 3·34 kg	Wt 3·40 kg

Shadow F1	F3	Rondel F2	Brabham F3
13″x 11″front	13″x 8″front	13″ x14″rear	13″x 10″ rear
Wt 3·63 kg	Wt 2·95 kg	Wt 3·80 kg	Wt 3·46 kg

Fig. 1.3 Cross sections of several MELMAG magnesium alloy racing wheels.

expected to survive with a rim thickness of no more than 2 mm. As with all components in a racing car, a high standard of reliability can never be achieved if the car is to stand any chance of winning.

Many Grand Prix cars today are fitted with the MELMAG wheel, invented by Gerry Watts of Magnesium Elektron Ltd of Swinton, Manchester. Cross sections of several MELMAG designs from Formula 3 to Formula 5000 are given in Fig. 1.3. The MELMAG wheel consists of two deep pressings in magnesium alloy ZM21. A disk of honeycomb foil, mounted on the central magnesium alloy hub attachment, is used as a spacer element. A tubular tension strap envelopes the central honeycomb spacer. The complete assembly is bonded together by a high-temperature adhesive. This sandwich construction of the wheel disk, in which a honeycomb spacer is used to separate two stressed outer skins, is an aerospace technique.

1.3.4 Wire wheels

The centre-lock wire wheel (see Fig. 1.4) is traditionally associated with vintage sports cars and racing cars, and for

Fig. 1.4 Centre lock wire wheels with octagonal hub caps.

those of us of advancing years the blood is still stirred by memories of split-seconds saved by the deft application of copper-headed hammers to eared hubcaps. We also remember the problems of fretting corrosion that could occur on the conical seatings and the continual attention demanded by the wire spokes. Only Ettore Bugatti ever made a wire wheel that was substantially lighter than a good design of steel wheel and for this he used a very large number of spokes of fine-gauge, high-tensile piano wire. The excellent brake-cooling afforded by wire wheels made them irreplaceable in the 1930s. Even with the 1937 Formula 1 Mercedes that transmitted about 600 b.h.p. to the road surface, wire wheels were still reliable when serviced with Teutonic regularity.

Centre-lock wire wheels are now only fitted as an 'optional extra' on a few sports cars. The weight-saving is negligible. The Morgan 4/4, for example, can be supplied with 5 × 15 centre-lock wire wheels of 8.75 kg weight. A replacement steel wheel would be no heavier, but the aesthetic appeal is undeniable and the type of driver who buys a Morgan usually prefers wire wheels.

The Wheel Division of Dunlop Ltd also supply chrome-plated bolt-on wire wheels. These again offer no saving in weight, but are a direct replacement for the standard 4- or 5-stud steel wheel. They are optional equipment on such sports and 'sporting' cars as the Datsun 280Z, the Opel Manta and the Ford Mustang.

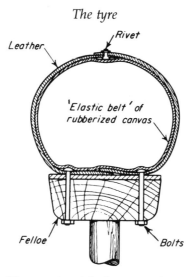

Fig. 1.5 Thomson's original pneumatic tyre of 1845.

1.4 The tyre

When the process of vulcanizing natural rubber became a commercial proposition (albeit a very poor product by modern standards), it was no great step to replace the steel-rimmed wheel by one carrying a solid rubber tyre. I myself have painful memories of solid tyres, still in use on commercial vehicles when I entered my teens. At 13 I drove a 4-ton Leyland lorry (illegally, of course) with cast-iron spoked wheels fitted with solid rubber tyres. The shocks transmitted up by the steering column by the Lancashire stone setts at 12 m.p.h. (the legal speed limit for a lorry) were almost as bad as the vibrations of a pneumatic road hammer. Even so, it was not the idea of comfort that led John Boyd Dunlop to make pneumatic tyres for the tricycle ridden by his 10-year-old son. He noted what a struggle the lad had to pedal his tricycle along a muddy lane. Fat-section pneumatic tyres, he reasoned, would not sink so readily into the soft mud and would not require as much physical effort from the rider. Dunlop made his first pneumatic tyre in 1888 and remained unaware for many years, even after a company had been formed to manufacture his tyres, that Robert William

Thomson had patented a pneumatic tyre in 1845 (see Fig. 1.5) and had even made measurements of the reduced tractive effort required to pull a light brougham over a paved road, a macadam road and a surface of crushed granite, when a change was made from wrought-iron tyres to pneumatic ones.

There was little interest in Thomson's invention at the time. No doubt horses would have appreciated this improvement, but horses are not as articulate as cyclists and tyres did not appear on the scene in any numbers until the end of the nineteenth century. Cyclists very soon established superior ride and reduced road resistance of the products marketed by the Pneumatic Tyre and Booth's Cycle Agency Ltd, of which J. B. Dunlop was a director. Since Dunlop's company* did not have a monopoly, there were many people working on the problems associated with these most unreliable early pneumatic tyres. It was the Michelin Co., in France, who were the first to market a pneumatic tyre for the new horseless carriages.

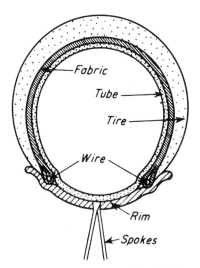

Fig. 1.6 C. K. Welch's rim and bead design of 1890.

*Later reregistered as Dunlop Pneumatic Tyre Co.

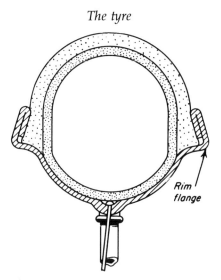

Fig. 1.7 William Bartlett's rim and bead design of 1890.

Two inventions that improved the reliability and convenience of the pneumatic tyre were registered in 1890. An Englishman, C. K. Welch, invented a *detachable* tyre incorporating high-tensile steel wires in what we now call the 'beaded edge' (see Fig. 1.6). Only 36 days later an American, William Bartlett, filed an invention (Fig. 1.7) with an improved beaded edge that was locked in position on the rim by the internal air pressure. Bartlett did not use steel wire in his beaded edge. By combining the two systems, we arrive at the beaded-edge construction used in 99% of modern tyres. The internal construction of a modern cross-ply (*bias-ply* in the USA) tyre is shown in Fig. 1.8.

1.4.1 Tyre construction

The early tyre makers decided to strengthen the walls of the tyre with a cotton–canvas reinforcement. This seemed a sensible decision at the time but it proved to be a disastrous mistake. By using both warp and weft interwoven in this manner, a chafing action resulted as the tyre section flexed at the road contact patch. This generated heat in the natural rubber compound and led to a rapid breakdown in the

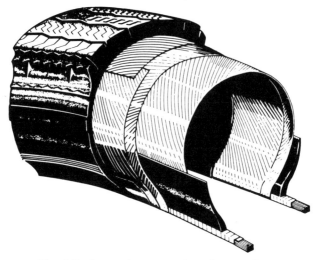

Fig. 1.8 Internal construction of cross-ply tyre.

reinforcement. The average life of the early car tyre was about 2000 miles.

In the 1920s the cross-ply construction was developed in which the tyre casing was built up by placing several plies of parallel cords inside the mould, the cords passing at a bias angle from the inside beaded edge to the outer beaded edge. Alternate plies, however, had the bias angle (relative to the circumferential centre line) alternate between a positive and a negative angle. The all-important difference from the earlier canvas reinforcement is that each separate sheet of ply has parallel cords which are coated completely with a rubber compound. Direct contact between adjacent cords is, therefore, avoided. The alternating angles between the bias plies and the circumferential centre line is usually between 20° and 30°. Large angles are chosen when comfort is the primary consideration. Racing tyres are made with an angle of about 20°, giving a harsh ride, but increased resistance to the centrifugal forces associated with racing speeds.

Until 1948 when Michelin introduced the 'X' series radial tyres, many tyre manufacturers believed there was little left for their research-and-development departments to do but to make minor refinements in construction techniques and to

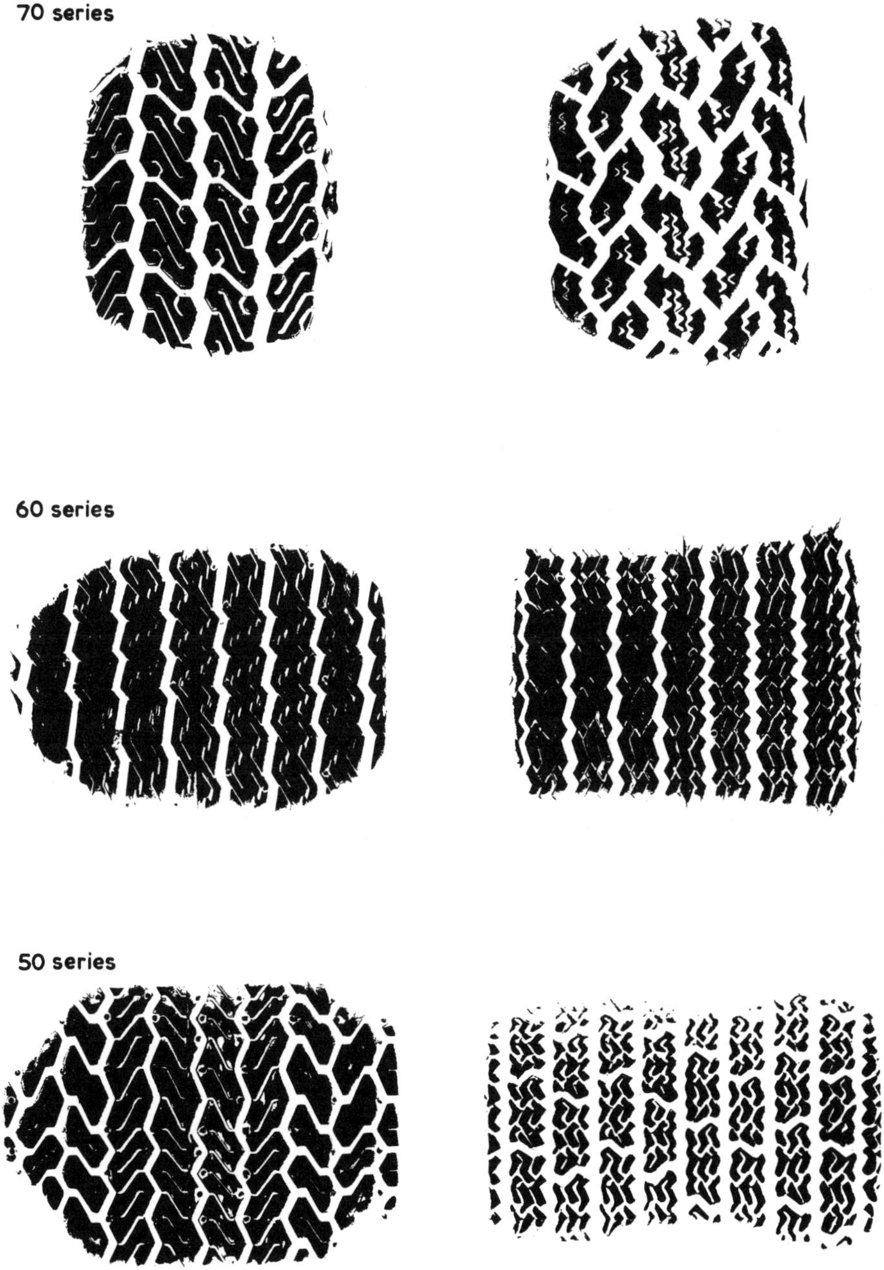

70 series

60 series

50 series

Fig. 1.9 Tyre footprints.

improve the grip and wear resistance of the tread compounds and patterns. They were well aware that the contact patch (or footprint) (see Fig. 1.9) suffers a distortion and a scrubbing action under cornering forces and, to a less extent, under acceleration and braking. Tread patterns and rubber compounds were developed to resist this action. In the new 'X' tyres Michelin changed the method of tyre construction to reduce the amount of distortion in the footprint zone and immediately doubled the life of car tyres using identical compounds. Michelin stiffened the zone behind the tread by bands of steel mesh. They also increased the flexibility of the side walls by increasing the cord bias angle to 90°; hence the name 'radial', since the textile cords passed radially from inner to outer beaded edge.

Since the Michelin patents expired many variants on the theme have emerged, but the common factor is the use of

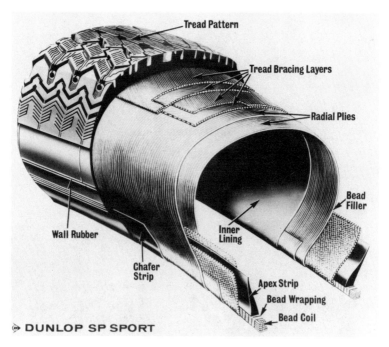

Fig. 1.10   Internal construction of steel-braced radial tyre.

radial cords. Some manufacturers still use steel mesh to stabilize the footprint, some use nylon belts sandwiched between steel belts, others use belts made only of synthetic textiles. A modern example is shown in Fig. 1.10.

A characteristic of the original 'X' tyre was a superior cornering power that had a tendency to fall rapidly as the limiting cornering speed was reached. With modern radial tyres the breakaway is more gradual and well within the control of the average driver.

With normal road speeds and moderate cornering modern radial tyres give mileages of 40–50 000 miles. This has been something of a mixed blessing to the tyre makers!

1.4.2 Skidding

In the 1920s when tread patterns of the type shown in Fig. 1.11 were in vogue, it was known that ribbed tyres gave long life while block patterns gave good resistance to skidding. Until makers like Dunlop developed their Cornering Force

Fig. 1.11 Tread patterns of the 1920s.

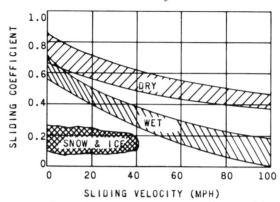

SLIDING VELOCITY (MPH)

Fig. 1.12 Effect of sliding velocity on coefficient of friction.

Machine in the 1950s tread patterns were developed and tested by a loosely co-ordinated system of road-testing and feedback of information from the customer. It was 1958 before the 1st International Skid Prevention Conference was held, and 1964 before the American Society for Testing Materials standardized a Standard Pavement Skid Test Tire in their specification ASTM-E17.

Early experiments using this tyre soon gave scientific proof of the obvious; the type and condition of the road surface is just as important as the tyre. On wet road surfaces (*pavements*, in the USA) the coefficient of friction could vary from almost zero to 0.7. On dry roads the variation was less drastic, being from 0.4 to 0.87. It will be seen from Fig. 1.12 that the coefficient falls with increase of speed. The Road Research Laboratory in the UK had already measured locked-wheel stopping distances on good road surfaces, such as dry clean concrete or hot-rolled asphalt with precoated chippings, of 30–40 ft from 30 m.p.h. This represents a coefficient of friction of about 0.8. One of the poor surfaces gave a stopping distance from 30 m.p.h. with locked wheels and in wet conditions of 495 ft. The coefficient of friction in this case would be very close to zero, confirming the measurements made on the Standard Pavement Skid-test Tire.

It is interesting to note that smooth tyres gave almost as good frictional coefficients as those with tread patterns when

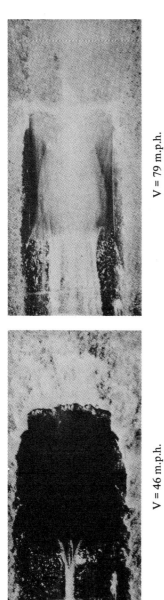

V = 46 m.p.h.

SMOOTH TREAD

V = 79 m.p.h.

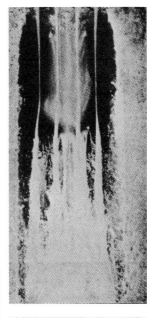

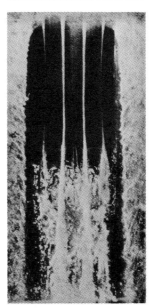

V = 46 m.p.h.

4- GROOVE RIB TREAD

V = 80 m.p.h.

Fig. 1.13 Aquaplaning: water depth, 0.5 mm (photographed from below through transparent road surface).

tested on good dry surfaces. To achieve maximum footprint area, *dry* racing tyres are completely bald. They are, of course, extremely dangerous on a wet road, since they behave like surfboards. At speeds above about 50 m.p.h. the buildup of water in front of the footprint acts like a wedge to force a film of water across the whole area of the footprint, thus destroying the grip on the road. This phenomenon is known as *aquaplaning* (*hydraplaning*, in the USA). All good modern tread patterns are designed to allow large volumes of water to pass from front to rear of the footprint. The influence of good footprint drainage is clearly illustrated in Fig. 1.13.

1.4.3   Dynamic footprint behaviour

Skidding, as discussed so far, is complete loss of directional control, but this does not mean that the tyres travel over the road surface at all other times like the flanged wheels of a train on a track. When torque is applied to accelerate or brake the vehicle, or when centrifugal force is resisted in a corner, slip always occurs between the tyre and the road surface. Fig. 1.14 shows how a wheel operates at a slip angle to create a

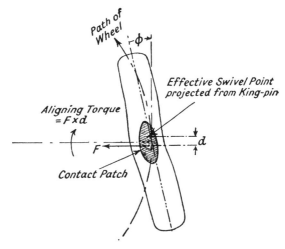

Fig. 1.14   Distortion of tyre as it approaches footprint zone (extent of distortion exaggerated).

cornering force. The slip angle increases if more cornering force is demanded. This can increase until a point is eventually reached where an increase in slip angle produces no increase in cornering force. This is the limit. It occurs at a slip angle between 10° and 14°, depending upon the particular tyre design and, of course, the road surface. Beyond this limit the cornering force decreases with increase in slip angle and, if the driver makes no steering correction to reduce the centrifugal force, the tyre begins to slide. This is a true skid.

The side thrust exerted by the footprint also varies with the load carried by the tyre, since the size of the contact patch varies with load. For every value of slip angle there is an optimum load and an optimum tyre pressure. A typical family of curves measured by a cornering force machine is shown in Fig. 1.15. With a change in tyre pressure, a different family of curves would result. Modern low-profile tyres are very

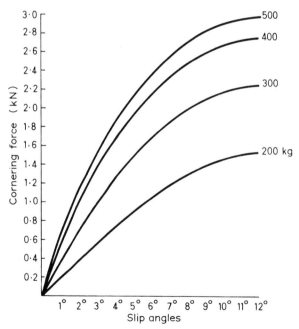

Fig. 1.15 Typical cornering force curves for modern radial.

sensitive to tyre pressure. If the pressure is too high, the
footprint area is reduced. If the pressure is below the
optimum, the footprint becomes concave. More load is carried
by the outside zone of the footprint, less by the central zone.
The right-hand 60 series and 50 series footprints shown in
Fig. 1.9 appear to be slightly underinflated.

1.4.4 Cornering power

Cornering power is defined as cornering force divided by the
slip angle required to generate the force. It is an important
parameter to all tyre designers, but is difficult to show in
graphical form since the family of curves usually falls so close
together. If we study Fig. 1.15, we see that this particular
tyre at a load of 300 kg has a cornering power of
0.31 kN/degree at a slip angle of 2 degrees; 0.305 kN/degree
at 4 degrees; 0.275 kN/degree at 6 degrees; and 0.245
kN/degree at 8 degrees. This is typical, with the cornering
power falling with increase in slip angle.

We have already mentioned how tyre pressures and vertical
load can influence cornering forces, and by inference
cornering power. The following factors also influence
cornering power:

Tyre profile. The tyre profile ratio is usually expressed as an
aspect ratio, this being the ratio of height to width of the tyre
cross section. It is variously described by the tyre makers as
'70 profile' or '70 series' when the height is 70% of the width.
Thus a 225/50 VR16 marking on a tyre indicates a tyre width
of 225 mm with a 50 profile on a 16-in diameter wheel and a
VR speed rating (up to 150 m.p.h.). The mixed units are an
indication that metrication is a slow process of integration
across many industries. About twenty-five years ago tyre
profiles had only moved from 100 to 85. The actual profile
(*not* profile ratio) of the footprint at this time was convex to
the road surface. Measurements of cornering power showed
that a small increase in tyre pressure above the maker's
recommended pressure would give an increase in cornering
power. Modern low-profile tyres present a flat profile to the
road surface when at the designed load and at the
recommended pressure. Very high cornering powers are
given by those with the more extreme aspect ratios and it is

almost impossible for cautious, elderly motorists like the present writer to make a car like the Porsche type 928 lose its grip on a good dry surface. The use of very wide tyres, however, does demand certain restrictions in the suspension system, since a change in camber angle of more than two or three degrees will lift the tread away from the road surface on one side and seriously reduce cornering power. Racing cars use aspect ratios as low as 35% to achieve quite phenomenal cornering powers. Such tyres are very sensitive to camber changes and tyre pressure and are a source of considerable worry to both driver and tyre technician.

Rim width. There is an optimum wheel rim width for any tyre section. If a change is made to tyres with a lower aspect ratio, it is advisable to change the wheels, too, since the use of wide tyres on narrow rims introduces the danger of destructive stresses in the side walls.

Camber. Wheel camber is the angle between the plane of rotation of the wheel and the vertical. It is conventionally

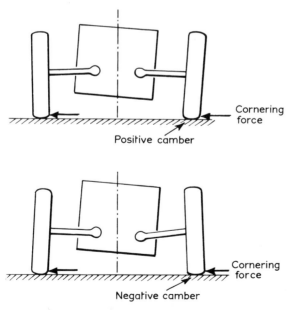

Fig. 1.16 Camber.

defined as positive when the top of the wheel leans towards
the cornering side force. This is illustrated in Fig. 1.16. As
already stated, with tyres of aspect ratio greater than 80% an
increase in cornering power is given when a wheel is run at a
negative camber. By extending the tread pattern well into the
side walls, racing motor cycles gain the ability to corner at
very high speeds, using what appear to be impossibly high
negative cambers. It is interesting to compare this with the
contrasting development in tyres for racing cars where
ultra-low aspect ratios demand the virtual elimination of
camber.

Load. Every tyre is designed for a specified load. There is a
certain flexibility in this parameter and the tyre pressure is
usually increased to restore the normal footprint when the
car is heavily loaded. A large increase above the recommended load, however, increases operating tempera-
tures. The safe operating speed is thus reduced on a heavily
loaded car.

Another aspect of load variation of interest to the designers
of high-performance cars is that excessive load transfer when
cornering at high slip angles reduces the ultimate cornering
power. For example, if we take our cornering force values
from Fig. 1.15, we find that a car designed with negligible
load transfer and with both inner and outer tyres loaded to
300 kg, would exert cornering forces of 1.97 kN from each
tyre at slip angles of 8°. If however the suspension geometry
permits roll to take place to the extent that the inner wheel
load is only 200 kg and the outer is increased to 400 kg,
greater slip angles will be required to exert the required total
cornering force of 3.92 kN. A value of 9.1 degrees would give
about 2.59 kN cornering force on the outer wheel and
1.38 kN on the inner. Load transfer, therefore, reduces
cornering power.

Traction, braking and acceleration. It has been shown
experimentally that with few exceptions a tyre footprint will
exert the same limiting force *in any direction*. Under braking or
acceleration, the limiting cornering forces are therefore
reduced. A close approximation to the limiting cornering force
in such cases is given by a vectorial combination of the two

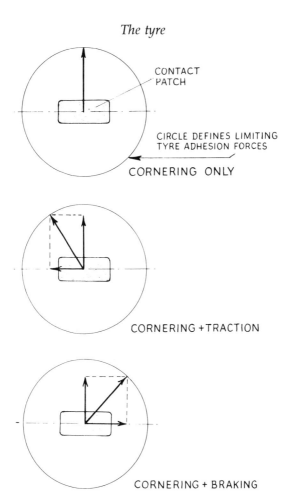

Fig. 1.17 Circle of Forces: how available cornering force is reduced during traction and braking.

forces (cornering plus traction, or cornering plus braking). The circle of forces shown in Fig. 1.17 demonstrates this.

Reference

[1] Setright, L. J. K. (1972), *Automobile Tyres*, Chapman & Hall, London.

2
Springs

2.1 Comfort and fatigue

The degree of discomfort we are prepared to accept from an automobile suspension depends largely on the vehicle and its purpose. A racing driver will endure vibrations that blur his vision and vertical accelerations that jar his spine, yet he only mentions such things to the team manager if he has noticed a subtle change from the norm, since such a change could indicate a flat spot on a tyre or an impending transmission failure. This same driver, however, will complain that his new sports car has a hard ride, even though by comparison with his racing car the sports car gives a ride like a feather-bed. In a sports car most drivers will accept a relatively hard ride as a necessary adjunct to precise high-speed handling and safe fast cornering. In a large saloon a similar ride would be considered intolerable.

2.2 Tolerance to vibration

The medical profession has been interested in the effects of vibration on the human body for many years. In particular, studies have been made to establish which particular frequencies are harmful and produce fatigue quickly in industrial workers. Raynaud's phenomenon, for example, is associated with high-frequency vibrations in the range 20–200 Hz from handheld power tools, while motion sickness is more likely to occur with very low-frequency vibrations in the range 0.1–0.5 Hz. As a broad guide, we can say that the human frame experiences fatigue most quickly when exposed to vibrations in the range 4–8 Hz. A working group of Technical Committee 108 of the International Organization for Standardization gave recommendations for the time-limits beyond which one can expect a worker to show signs of 'fatigue-decreased proficiency'. Their recommended time-limits for vibrations at an RMS acceleration level of 1.0 m/s² (approx. 0.1 gravity) were 1.5 h at a frequency band of 4–8 Hz, increasing to 4 h at 1.0 Hz. Obviously we should aim to provide a suspension system that confines the major vibration frequencies very close to 1.0 Hz, avoiding if possible any prolonged subjection of the occupants to frequencies as low as 0.5 Hz.

2.3 Spring design

2.3.1 The simple undamped spring

A simple spring system is shown in Fig. 2.1. In the mean position the force exerted by the coil spring P exactly balances the weight (or gravitational force) W of the mass M. If the mass is displaced by distance x, the spring force will increase to $P + kx$, where k is the stiffness or rate of the spring, i.e. the force per unit displacement.

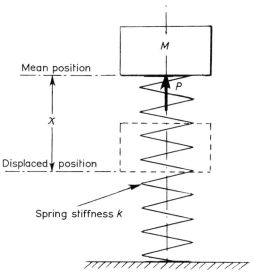

Fig. 2.1 Simple spring system.

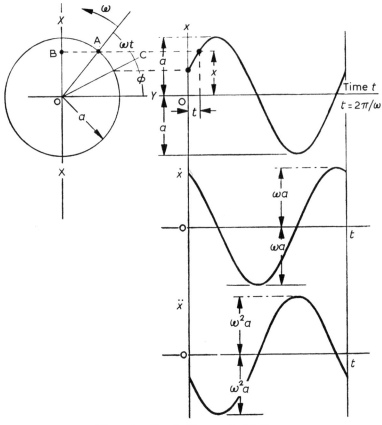

Fig. 2.2 Simple Harmonic Motion.

In the displaced position the body of mass M, if released, will accelerate towards the mean position at a value of $\ddot{x}$. Since x is considered positive in a downward direction, this acceleration will also be positive. Conversely, if the body is displaced in an upward direction by an amount $-x$, the acceleration will be in the opposite direction. This system, therefore, constitutes a Simple Harmonic Motion and the periodic time for such a motion is:

$$T = \frac{2\pi}{\omega} = 2\pi\sqrt{\frac{M}{k}} . \qquad (2.1)$$

ω is the angular speed in Circular Harmonic Motion, as illustrated in Fig. 2.2. This equation can be expressed in words as:

$$\text{Periodic time} = 2\pi\sqrt{\frac{\text{Mass of body}}{\text{Stiffness of spring}}}. \qquad (2.2)$$

Alternatively, if we take the deflection of the spring under the weight of the body as d (the static deflection):

$$T = 2\pi\sqrt{\frac{d}{g}} \qquad (2.3)$$

where g, in this case, is the acceleration due to gravity.

2.3.2 Body accelerations

Until road authorities provide road surfaces as smooth as a billiard-table, automobiles will continue to need springs. Let us take the simple case of a car travelling at speed that encounters a 50-mm step in the road level, the kind of step one could easily encounter when a new road surface is being laid. For simplicity we will neglect the spring in the tyres, the behaviour of the spring dampers and the springs in the seats.

The sudden compression of the front springs of the car by a 50-mm step in the road will not produce an instant lift of 50 mm in the car body. This would be a shattering experience to both car and occupants. What actually happens is a compression of the two front springs by 50 mm. The springs then begin to release this stored energy by lifting the front end of the body. The rate at which this occurs is controlled by the natural frequency of the springs. The maximum acceleration transmitted to the body is given by a rearrangement of the Equation (2.3):

$$a = d\left(\frac{2\pi}{T}\right)^2. \qquad (2.4)$$

Taking a typical front spring frequency of 1.25 Hz, i.e. $T = 0.8$ s:

$$a = 0.05 \times \left(\frac{2\pi}{0.8}\right)^2$$
$$= 3.08 \text{ m/s}^2$$
$$\approx 0.31 \text{ gravity.}$$

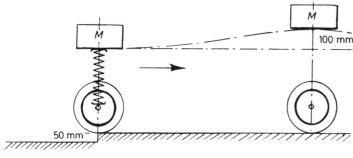

Fig. 2.3

This maximum acceleration to which the occupants will be subjected will occur at a quarter-oscillation of the springs after the front wheels strike the 50-mm step, i.e. after 0.2 s. If the car is travelling at 50 m.p.h. (22.4 m/s), the occupants will feel the maximum acceleration about 4.5 m after the wheels strike the step. With the undamped springs the initial compression of 50 mm will lift the front end of the body a total distance of 100 mm, since the springs will overshoot the equilibrium position by another 50 mm (see Fig. 2.3).

Effective damping is, therefore, essential if the car is not to proceed along the road like a yo-yo. For a given spring rate, good damping does however carry the penalty of an increase in maximum acceleration. On the other hand, good damping does permit the use of a softer suspension and the overall effect is most certainly beneficial.

If we doubled the spring rate of the example above, i.e. using springs with a frequency of 2.5 Hz, as one might find on a vintage sports car, the maximum acceleration would become:

$$a = 0.05 \times \left(\frac{2\pi}{0.4}\right)^2$$
$$= 12.4 \text{ m/s}^2$$
$$\approx 1.26 \text{ gravity.}$$

With high-pressure narrow tyres and very little help from vintage-type upholstery, the inevitable result would be that the occupants would momentarily lose contact with their seats. This explains why drivers at Brooklands used body belts to support their abdominal muscles against the hammering inflicted by the irregularities of the concrete oval.

2.3.3   Spring rates

Springs can take a multitude of shapes: parallel coils, conical coils, torsion bars, spirals, disks, single leafs, laminated leafs; there seems to be no end to the possibilities. One need not be confined to metallic materials. They can be made of rubber or take the form of a gas container compressed by a piston or diaphragm. Today the most popular spring in use on automobiles is the coil spring with coils of identical diameter and spacing. It is also convenient and certainly inexpensive to use a constant gauge from top to bottom. Such a spring usually gives a close approximation to a constant spring rate over its working range. If we compress it 10 mm and it exerts a force of F Newtons, compression to 100 mm will give a restoring force of $10F$ Newtons. One can, of course, vary the spring gauge or the diameter of the coils to give a variable-rate spring, but our mathematical analysis would, at this stage of our study, become very long and tedious if we introduced such complexity.

There are certain physical limitations to the use of very low spring rates. The lower the spring rate the greater the initial deflection (the *static deflection*) under the weight of the body and, of equal importance, the greater will be the total travel

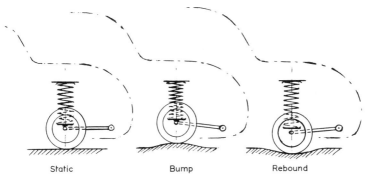

Static   Bump   Rebound

Fig. 2.4 Bump and rebound.

of the wheel in bump and rebound (*jounce*, in the USA) (see Fig.2.4).

Let us consider a case for which the sprung mass M, is 1200 kg, which with a 50/50 weight distribution front to rear gives a load of 300 kg per wheel. From Equation (2.1), we derive:

$$k = \frac{M}{4} \left(\frac{2\pi}{T} \right)^2$$

or, since $T = 1/f$ (where f is the frequency in Hz):

$$k = \frac{M}{4} (2\pi f)^2 \text{ N/m.} \qquad (2.5)$$

Table 2.1 demonstrates how an increase in spring natural frequency increases the spring rate and decreases the static deflection. This table helps to highlight our problem. If we want a comfortable ride, the spring frequency must be low, preferably below 1.5 Hz. To achieve this, the suspension system and its attendant linkages must permit large wheel movements, often as high as 250 mm (10 in). When modern tyre designs with low-profile ratios demand suspension systems that give negligible change in camber angle, we begin to see how hedged around with conflicting requirements we have become. It will take several chapters before we are able to show how a few contemporary suspension engineers have achieved satisfactory compromises between these conflicting demands.

Table 2.1 *Spring rates and static deflections under 300 kg sprung mass*

Frequency f (Hz)	Spring rate k		Static deflection d	
	(N/mm)	(lbf/in)	(mm)	(in)
1.00	11.8	67	248	9.8
1.25	18.5	105	159	6.3
1.50	26.7	151	110	4.3
1.75	36.3	206	81	3.2
2.00	47.3	269	62	2.4
2.50	74.0	422	40	1.6

2.4 Wheel contact

Wheel contact with the road surface is obviously desirable at all times, yet we know from experience that it is not all that uncommon for one or even both wheels at front or rear of the car to leave the ground. In some of the more spectacular rallies the cars only seem to spend about half the time with all four wheels on the ground.

What are the factors, then, that help us to maintain the tyres in contact with the ground? For simplicity, the ripple is considered to be sinusoidal and the movements of the two front wheels to be in phase. Again to simplify matters the mass of the front part of the body, $2M_f$ is considered to be capable of independent movement in relation to the rear. We can, therefore, treat a single front suspension system as if it were a pogostick with a single mass M_f mounted on one wheel (see Fig. 2.5). The unsprung mass, i.e. the front wheel, approximately one-half of the coil spring and one-half the mass of the suspension links, is m_f.

Let us take the case of a spring of frequency $f = 1.5\,\text{Hz}$ and a ripple depth $x = 150\,\text{mm}$. From Table 2.1, we see this depth exceeds the static deflection of the spring (110 mm). The tyre contact patch will, therefore, be completely unloaded at the bottom of the ripple. The unsprung mass has inertia and will cause the wheel to overshoot this point, but not by

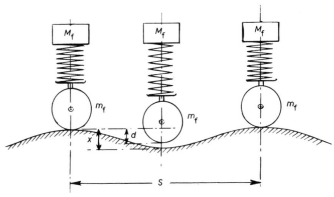

Fig. 2.5 Loss of contact in a ripple.

as much as 40 mm. Therefore, we can predict that a ripple depth of about 150 mm is the limit for such a spring system, if wheel contact is to be maintained.

If we soften the spring suspension to a frequency of 1.25 Hz, the static spring deflection is increased to 159 mm. With this new spring the tyre will remain in full contact when traversing the length of the ripple, even though the load on the footprint will be reduced at the bottom of the ripple.

We have shown that our pogostick system can be designed to maintain reasonable surface contact with a ripple depth of 150 mm (6 in) with a relatively low frequency. There must, however, be a lower limit to the length of ripple that can be traversed at a given speed. In this analysis it is assumed that the car is travelling at such a high speed that the inertia of the relatively massive sprung body prevents any appreciable vertical movement. The conditions leading to resonance bouncing of the body will be discussed later. The natural periodicity of the spring under the action of the unsprung mass is not the same as that under the action of the sprung mass. From Equation (2.1), we see that $T \propto \sqrt{M}$. Thus, with a natural period of T under mass M_f, $t = T\sqrt{(m_f/M_f)}$ when the spring is oscillating under the smaller mass m_f.

If we take the limiting ripple length as s metres and the car velocity as v metres per second, the limiting ripple length is given by:

$$t = T\sqrt{\frac{m_f}{M_f}} = \frac{s}{v} .$$

Therefore

$$s = vT\sqrt{\frac{m_f}{M_f}}$$

$$= \frac{v}{f}\sqrt{\frac{m_f}{M_f}} \qquad (2.6)$$

where f is the natural frequency in Hertz under sprung mass M_f.

Table 2.2 has been calculated for a range of values of m/M and for a range of spring frequencies. The speed in all cases is 100 k.p.h. (62 m.p.h.).

Table 2.2 *Limiting ripple length for road contact at 100 k.p.h.*

Spring frequency f (Hz)	Limiting ripple length for road contact at 100 k.p.h. (62 m.p.h.) (m)		
	$m/M = 1/4$	$m/M = 1/6$	$m/M = 1/10$
1.0	13.9	11.4	8.8
1.5	9.2	7.6	5.9
2.0	6.9	5.7	4.4
2.5	5.6	4.5	3.5

A close study of Tables 2.1 and 2.2 shows that a low spring frequency helps the wheels to follow the contours of road ripples at speeds of 100 k.p.h., but only if the ripples are spaced at a greater distance than 9–14 m.

On the other hand, the use of a high spring frequency, such as 2.5 Hz, will reduce these limits to 3.5–5.5 m, depending upon the unsprung-to-sprung mass ratio, but the limiting depth of the ripple will be correspondingly reduced. The choice of spring frequency is, therefore, not easy. Once again, we are faced with that unattractive word 'compromise'. Many modern designers are turning to rising rate suspensions in an attempt to conquer this problem. Such springs give low frequencies under small wheel movements and higher frequencies under large movements.

2.5 Movement of the sprung mass

Keeping the wheels on the ground is one design consideration, but no less important in a passenger-carrying vehicle, is the movement of the sprung mass, particularly since you or I, or even our mother-in-law, could be part of it. Returning to our pogostick concept, we can apply Newton's Second Law of Motion. The force imparted by the wheel to the base of the spring will thus produce an acceleration of the sprung mass. Since this force is also related to the mass and acceleration of the unsprung mass, we can state that:

$$F = ma = MA$$

where a = the acceleration of the unsprung mass, and A = the acceleration of the sprung mass.

As a general case, we can state that any road disturbance producing an acceleration a in the unsprung mass, will transmit an acceleration in the sprung mass of value $A = a(m/M)$. For maximum comfort and, of course, to reduce the shock loads on the body/chassis components to a minimum, the ratio m/M must be as low as possible.

This general case applies to the typical road contour composed of random surface irregularities. The case of a regular series of undulations, such as one meets on unmade roads in undeveloped countries or on the 'washboard' dirt roads in the less-populated areas in the USA, is a special case that calls for special treatment. The evenly spaced ridges in a washboard road are produced by the natural frequency of the suspension system, since the driver adopts a speed that produces the least body movement. In theory a car travelling at the optimum speed with *identical* frequencies at both ends could float along with negligible vertical accelerations transmitted to the passengers. In practice a typical car would have a mean frequency of about 1.4 Hz (1.3 Hz at the front and 1.5 Hz at the rear). If we take this mean value of 1.4 Hz and a mean of m/M of 1/10, the most comfortable value of v can be calculated from Equation (2.6):

$$s = \frac{v}{f}\sqrt{\frac{m}{M}}.$$

Therefore

$$v = sf\sqrt{\frac{M}{m}}.$$

If s is taken as 5 m:

$$v = 5 \times 1.4\sqrt{10}$$
$$= 22 \text{ m/s}$$
$$= 80 \text{ k.p.h. (49 m.p.h.).}$$

A car with a smaller m/M ratio, say 1/12, and with the same mean spring frequency, would travel most comfortably at a speed of about 87 k.p.h. (54 m.p.h.).

Vintage sports cars with mean spring frequencies of about 2.0 Hz or even higher would be at a great disadvantage in such circumstances, since the optimum speed for comfort would be much higher. Moreover, softly sprung modern cars would have dug troughs in the road to a depth of at least 80 mm. The vintage sports car would be bumping on its suspension stops, if the driver attempted to get 'into the groove'. Naturally if only cars with stiff suspensions ever used this road, the problem would not arise.

2.5.1 Body resonance

In the previous example, we considered the case of resonance of the suspension under the action of the *unsprung* mass. Resonance of the suspension under the *sprung* mass can occur and with no damping or inadequate damping this can be very disconcerting. Let us consider the case of a car with an undamped suspension system with a mean natural frequency of 1.4 Hz which encounters a series of ripples with a wavelength of 10 m.

At a velocity of $10 \times 1.4 = 14$ m/s = 50 k.p.h. (31 m.p.h.) the forcing frequency induced by the regular road ripples would exactly coincide with the natural suspension frequency

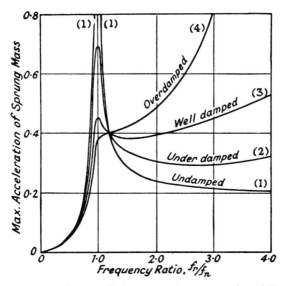

Fig. 2.6 Damping characteristics: note resonance when f_r/f_n is unity.

under the action of the sprung mass. Fig. 2.6 shows what happens on a typical undamped suspension system when this occurs. The undamped curve is indicated by the number 1. Naturally on a washboard road the driver would accelerate through the critical speed that caused body resonance. The profile of a typical road surface, however, presents a wide gamut of forcing frequencies and at any chosen cruising speed a car with undamped springing would quite frequently strike stretches of road that induced resonance.

Optimum damping is usually a compromise, a compromise that is often influenced by the tester's individual preference. Competition vehicles are sometimes fitted with adjustable dampers and are set up to meet the requirements of the particular driver. The contribution of the damper to suspension design is of primary importance and will be discussed fully in Chapter 7

2.6 The spring in the tyre

We have already commented on the effectiveness of the pneumatic tyre in isolating us from minor road surface irregularities. Even a road made of flat-topped paving stones, so popular in the larger towns and cities before the First World War, can be traversed in comfort with modern low-pressure tyres. At a speed of 50 k.p.h. (31 m.p.h.) stone setts of 6-in width would impart a vibration of about 90 Hz. If the setts are well laid with no more than 10 mm (0.4 in) variation in height between neighbouring setts, the flexing of the contact patch and the side walls will almost completely absorb the vibrations.

The complex construction of the modern tyre makes the application of mathematics to the flexural behaviour of tyres over a range of frequencies a daunting challenge. Even so, better men that the present writer have tackled the problem and Drs Overton, Mills and Ashley carried out a series of experiments at Birmingham University [1] in which they measured the dynamic behaviour of both cross-ply and radial tyres (non-rolling), successfully constructing a mathematical model that gives an accurate prediction of the experimental results.

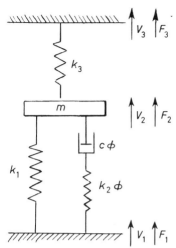

Fig. 2.7 Mathematical model used in Birmingham University experiments.

Distortion of the tyre carcase under cornering, acceleration or braking forces, could of course modify their basic concepts, but the major factors neglected by the fact that the measurements were made on non-rolling tyres are: (a) centrifugal expansion of the tyre carcase, and (b) the mobile nature of the footprint compression. The first factor would only make their predictions inaccurate at high speeds, but the second factor is a valid source of error since the continuous process of establishing a new footprint probably increases the effective tyre stiffness. It does not mean, though, that the critical vibration modes measured in these experiments are far from the truth.

The simple suspension model chosen by Overton, Mills and Ashley is shown in Fig. 2.7:

V_3 = velocity imparted to sprung mass;
F_3 = force imparted to sprung mass;
K_3 = spring rate;
m = unsprung mass;
V_2 = velocity of unsprung mass;
F_2 = force given to unsprung mass (at wheel hub).

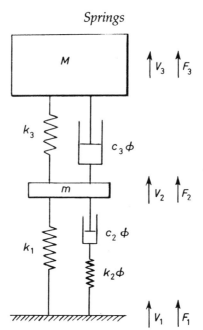

Fig. 2.8 Mathematical model for complete suspension system.

The 'springiness' of the tyre is represented by two springs. The first K_1 is undamped and can be regarded as the simple compression of the air in the tyre. The second K_2 represents the characteristic behaviour of the tyre carcase. Modern synthetic tyres exhibit a high degree of *hysteresis*. Expressed simply, this means that the material is not perfectly elastic and for every unit of kinetic energy absorbed by the rubber a proportion is converted into heat energy.

It is unfortunate that many synthetic rubber mixes that grip the road so well (high-μ compounds) are also high-hysteresis compounds. When taken to extremes the use of high-hysteresis compounds creates problems for racing team managers, since such tyres give excellent grip in the wet but overheat in the dry. Modern road tyres are a reasonable compromise, in that they give a good grip in the wet but do not overheat in the dry. The degree of hysteresis is still enough to give a fair measure of damping, and this damping action increases with increase of frequency. Hence the

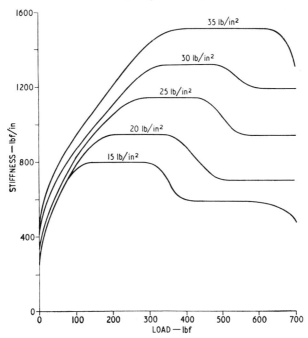

Fig. 2.9 Variation of stiffness with load and pressure for 5.60 × 13 cross-ply tyre.

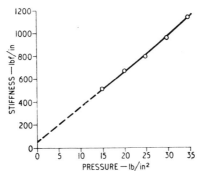

Fig. 2.10 Variation of stiffness with pressure for 165 × 13 radial tyre.

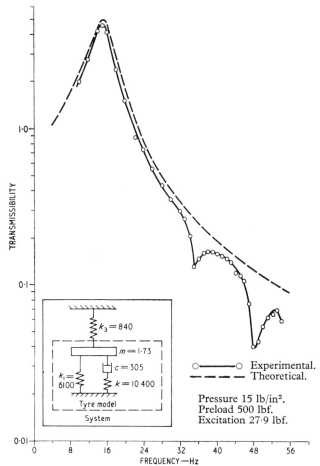

Fig. 2.11   Motion transmissibility across 5.60 × 13 cross-ply tyre.

addition of parameter ϕ, frequency, to the proposed viscoelastic model used in the Birmingham University study.

The Overton, Mills and Ashley model omits the suspension damper, since this does not enter into the resonance characteristics of the tyre. The complete suspension system (one wheel only) is represented more accurately in Fig. 2.8. The damping coefficient C_3 is expressed as N/ms. This

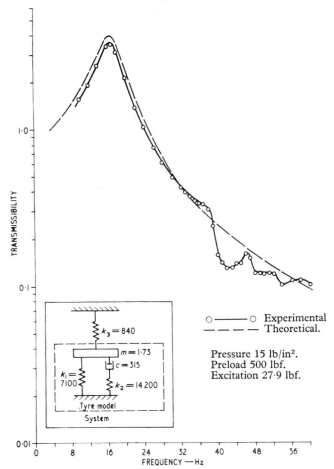

Fig. 2.12 Motion transmissibility across 163 × 13 radial tyre.

assumes that the damping coefficient increases directly with frequency.

Figs. 2.9 and 2.10 show typical measurements of tyre stiffness, or spring rate, plotted against load for a 5.60 × 13 cross-ply and a 165 × 13 radial tyre. The cross-ply shows a complex variation in stiffness. The radial tyre, however, shows a proportional variation with pressure. This is an early design of a steel-braced tyre with very flexible side walls. More complex modern radials using mixed textile bracing materials will probably show some variation with load.

2.6.1 Motion transmissibility

Motion transmissibility is defined by Overton, Mills and Ashley [1] as the ratio V_2/V_1, i.e. the ratio of wheel hub velocity to footprint velocity. Figs. 2.11 and 2.12 show both experimental and theoretical values for the two tyres at a tyre pressure of 15 lbf/in^2 (1.03 bar).

These results show that a tyre is a very effective vibration-absorber at frequencies above 40 Hz. Road impulses from a very low frequency up to about 20 Hz are transmitted unimpaired; in fact, at some resonance frequency about 16 Hz the velocity V_1 at the footprint becomes $V_2 \approx 4V_1$ at the wheel hub. This resonance frequency varies with tyre dimensions, inflation pressure and internal construction, but falls inside a range of 12–22 Hz.

At a speed of 50 k.p.h. (31 m.p.h.) a frequency of 16 Hz would be given by a sinusoidal wave formation in the road surface of 0.9 m (3 ft). This could be the cause of the 'boulevard bounce' experienced in the USA, where the popular fat tyres bounce so readily at typical city traffic speeds.

References

[1] Overton, J. A., Mills, B. and Ashley, C. (1969–70), *Proc. Inst. Mech. Engrs*, **184**, part 2a.

[2] Carlson, H. (1973), *Spring Designer's Handbook* (Vol I, Mechanical Engineering), Dekker, New York.

3
Suspension Principles

3.1 Coupled suspension

In Chapter 2 we concentrated on the behaviour of the front end of a car under vertical accelerations produced by irregularities in the road surface. For simplicity, we considered the front end of the car as a discrete mass supported on a main spring and an auxiliary spring (the tyre). If we omit such rare examples as the Panther 6, which has four steered wheels at the front, we can define an automobile within the remit of this book as a single mass supported on four wheels, and it is the behaviour of this integrated sprung mass on its four suspension systems that is our concern in this chapter.

In practice the most complex patterns of behaviour can occur and at any moment in time all four wheels can be moving up or down at differing frequencies, through different amplitudes and with phase differences between frequencies. With such formidable behaviour patterns it is not surprising that many car manufacturers with extensive test and development facilities still fail to strike the right balance by the time the car goes into production and find it necessary to modify the suspension design at a later date.

The large manufacturers have accumulated a vast pool of data on suspension design and are able to construct an accurate mathematical model before they design the suspension system for a new product. Even so, the suspension engineer can not achieve perfection using conventional suspension techniques. He can decide to sacrifice a measure of comfort to improve cornering, or he can decide that straight-ahead passenger comfort is his major priority as in a vehicle designed to convey expensive cut-glassware or a president, trusting that the vehicle will be driven round all bends with respectful decorum.

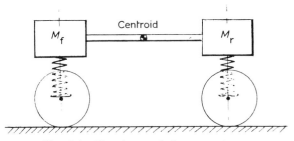

Fig. 3.1 Simple coupled suspensions.

3.2 Bouncing and pitching

Side-to-side suspension interactions will be neglected in this chapter. They are of course a major factor in cornering behaviour, but the phenomena of bouncing and pitching are experienced in the main when travelling in a straight line. In this chapter the pogostick of the last chapter has been replaced by two pogosticks connected by a rigid bar as shown in Fig. 3.1.

Pure bounce will occur if the front and rear sprung masses are equal, the front and rear springs have identical frequencies and identical rates and are in phase. We could design a car to have a 50/50 weight distribution and identical springs all round, but the phasing of the ripples in the road surface are beyond our control. It is inevitable that a degree of rocking or seesawing, or *pitching* as it is known in this context, must often occur. The inertia of the sprung mass will resist this pitching movement and the disposition of the major components, i.e. the engine, transmission, fuel tank, passengers, etc., that make up the sprung mass contribute to the resistance exerted by the sprung mass in opposition to the

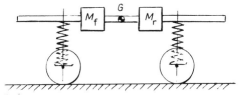

Fig. 3.2 Low polar moment of inertia.

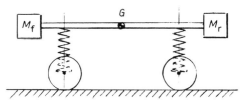

Fig. 3.3 High polar moment of inertia.

pitching moment. If the major masses of the body components tend to be near the centroid, as in Fig. 3.2, the sprung mass offers less resistance to pitching than a body of the same total mass that has the more massive components situated at a greater distance from the centroid, as in Fig. 3.3.

Expressed technically, we say that the second case has a larger *polar moment of inertia*. A moment of inertia is a moment of the second order, i.e. a moment in which each small unit of mass δm is multiplied by the square of its distance from a chosen axis:

$$I = \Sigma \, r^2 \, \delta m$$
$$= \int r^2 \, dm. \tag{3.1}$$

The axis, in this case, passes through the centroid. In this book we are interested in two polar moments of inertia. In a study of pitching the axis is horizontal (see Fig. 3.4). When we study the case of a car making rapid changes in direction, i.e. a racing car passing through a chicane, the axis of the relevant polar moment of inertia will be vertical.

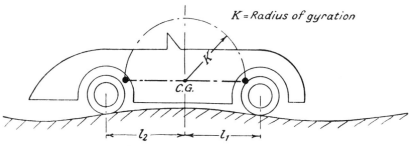

Fig. 3.4 Radius of gyration.

If the total sprung mass in Fig. 3.4 is M and we choose a value of K such that

$$MK^2 = \int r^2 dm \qquad (3.2)$$

the quantity K is called the *radius of gyration* of the mass about the axis. From Chapter 2 the period in bounce is given by

$$T = 2\pi\sqrt{\frac{M}{k}}. \qquad (3.3)$$

The period in pitching is influenced by the relative values of K and the distances l_1 and l_2 between the wheel centres and the centroid:

$$T' = 2\pi\sqrt{\left(\frac{M}{k} \times \frac{K^2}{l_1 \times l_2}\right)}. \qquad (3.4)$$

If $l_1 = l_2 = K$ and the front and rear spring rates K are equal, $T' = T$.

Although many modern cars have the centroid well forward, the value of $K^2/(l_1 \times l_2)$ is often close to unity. This is more by chance than by deliberate design.

3.3 Suspension theory

When a rigid body, such as a car body, is mounted on springs at both ends, and the front springs are subjected to vertical forces through the front wheels, oscillations are produced in both front and rear springs. A similar action occurs when the vertical forces are applied to the rear wheels. One of the earliest mathematical treatments of this reaction between front and rear suspensions was presented by Professor J. J. Guest [1].

For simplification, lateral forces are not considered. Moreover, it is also assumed that both front wheels are subjected to identical vertical forces simultaneously. The same conditions of identical forces in phase also apply to the rear springs. The mathematical model is, therefore, identical to our example of two pogosticks connected by a rigid bar, as shown in Fig. 3.1. Professor Guest's model is shown at the

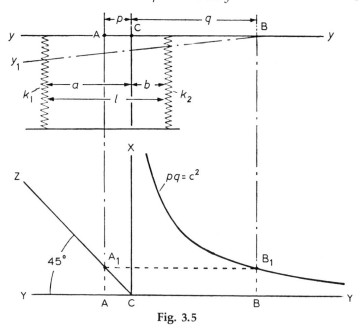

Fig. 3.5

top of Fig. 3.5. If a vertical force is applied at the front end of
the beam, the line XY which represents the centre line of the
body, will adopt an inclined position $X'Y'$, rotating about
point B. If a vertical force is applied at the rear, then rotation
will be about point A. Any pair of points that possesses this
relationship can be shown to satisfy the equation $pq = ab$. The
dimensions a and b are established by the relative rates of the
front and rear springs k_1 and k_2, since $k_1a = k_2b$. The spring
centre C is, therefore, the point of balance and a vertical force
applied at C will lift the beam with no tendency to tilt.

Points A and B are called *elastically conjugate points* by
Professor Guest. Since any value of p can be chosen, the
corresponding value of q is established by the relationship
$pq = ab$. Therefore, there is an infinite number of such pairs of
points. A geometric construction to help determine the pairs
of points is given in the lower portion of Fig. 3.5. The
hyperbola $pq = c^2$ is drawn relative to the axes XCY. For any
chosen value of P, the point A is projected vertically
downwards to meet the 45-degree line CZ at A. Projecting

horizontally then gives the corresponding value of B_1 on the hyperbola line. This can then be projected vertically upwards to give the other *elastically conjugate point* B.

It will be seen that as A approaches C, point B moves to the right and will move to infinity when A and C coincide. Similarly, as B tends to C, A will also tend to infinity.

The next stage of the Guest construction is to establish the *dynamically conjugate points* of the sprung mass. The moment of inertia of the sprung mass about the centroid is MK^2. Fig. 3.6 shows how the sprung mass M for which $I_g = MK^2$ can be represented by discrete masses m_1 and m_2 at distances r and s from the centroid G. To satisfy the equation $rs = K^2$, the following conditions must be met:

$$m_1 + m_2 = M$$
$$m_1 r = m_2 s$$
$$m_1 r^2 + m_2 s^2 = MK^2.$$

As in the case of the elastically conjugate points, the quantity r is chosen arbitrarily and the corresponding value

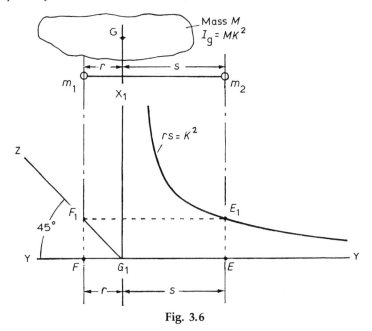

Fig. 3.6

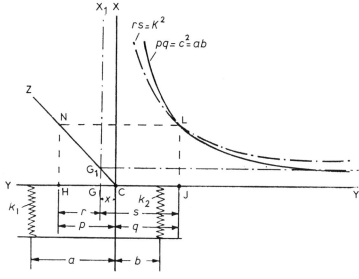

Fig. 3.7

of s is found from the relationship $rs = K^2$. Again there is an infinite number of pairs of dynamically conjugate points. The geometric construction is shown in Fig. 3.6.

The final stage is to combine the two systems. This will give us the *double conjugate points* that will satisfy both equations. This is shown geometrically in Fig. 3.7. Points H and J, projected from N and L (L being the point where the two hyperbolas cross) are points that are elastically and dynamically conjugate. These points are called by Professor Guest the double conjugate points; there is only one pair.

3.3.1 Special cases

When G and C coincide, but c^2 is not equal to K^2, the point of intersection of the two hyperbolas moves to infinity. This can be seen from Fig. 3.7. Point J is, therefore, also at infinity and point H coincides with G and C.

The other special case would occur when G and C coincide and $c^2 = K^2$. In this case the two hyperbolas also coincide and the double conjugate points are virtually indeterminate. It would be an interesting experiment for the inquiring student to observe how such a model behaves in the laboratory. (The writer admits he has always ignored this special case!)

To calculate the double conjugate points H and J, we have the following information (see Fig. 3.8):

$$pq = c^2 = ab. \tag{3.5}$$

$$rs = K^2. \tag{3.6}$$

$$p = r + x. \tag{3.7}$$

$$q = s - x. \tag{3.8}$$

For any given case, we know the suspension rates k_1 and k_2 and W, the wheelbase, which equals $a + b$. From this we know:

$$a = \frac{k_2}{k_1 + k_2}.$$

$$b = \frac{k_1}{k_1 + k_2}.$$

K^2 is a quantity that can be estimated by the student or can be measured accurately on special apparatus available to manufacturers or at certain research establishments such as the Cranfield Institute of Technology.

G can also be estimated in the drawing office (using such computer facilities as may be available) or, if the vehicle exists, by direct measurement of the axle weights on a weighbridge.

From the positions of C and G we know the value of x. Substituting for p and q in Equation (3.5), Equations (3.7) and (3.8) give us:

$$(r + x)(s - x) = c^2 \tag{3.9}$$

and substituting for s from Equation (3.6), we have:

$$(r + x)\left(\frac{K^2}{r} - x\right) = c^2$$

or $r^2x + r(x^2 + c^2 - K^2) - K^2x = 0$.

Hence:

$$r = \frac{K^2 - x^2 - c^2}{2x} \pm \frac{\sqrt{\{(K^2 - x^2 - c^2)^2 + 4K^2x^2\}}}{2x}. \tag{3.10}$$

This gives two values for r. The negative root is, of course, imaginary. There are also two values of s, a positive and a negative. s can be found from Equation (3.6) by substituting the value of r. It will be found that the negative value of s equals the positive value of r and vice versa. The two double conjugate points H and J are now known.

3.3.2 Properties of double conjugate points

The essential property of double conjugate points is that a force applied at one of them *will produce no motion at the other*. These independent motions are not pure bounce, however. Each end of the beam (or car body) makes an angular oscillation about its own conjugate centre, i.e. the front end about J and the rear about H. Unfortunately, if we try to approach pure bounce at the front end of the car by increasing the value of s, we increase the angularity of the motion at the rear by a corresponding decrease in the value of r. When road disturbances produce pitch, this additional angularity only adds to the discomfort of the rear passengers. Despite the help from headrests and reclining seats, the human frame is much less disturbed by bounce in moderation than by pitching. The frequency of the pitching is, of course, critical. A rocking chair can be soothing, but only when the frequency is low. A rapid rocking motion, i.e. of 2 Hz or more, tends to put a strain on the neck muscles. A ride in a short-wheelbase Jeep or Land Rover will illustrate this defect.

With conventional springing, the designer soon finds his options very limited. An analysis of several modern suspensions shows that many suspension engineers have chosen double conjugate points that coincide with the wheel centres. In this way bounce at either end of the car does not produce any bounce movement at the other. Spring rates at front and rear are usually chosen to give bounce frequencies that differ by at least 10%. Since the position of the centroid changes appreciably from the 'one-up' to the 'four-up' load, it is impossible to make the double conjugate points fall precisely over the wheel centres for every possible loading.

3.3.3 Spring frequencies about double conjugate points

When the front end of the car shown as a simple mathematical model in Fig. 3.8 pivots about its conjugate point J, the

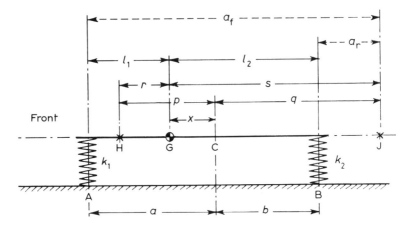

Fig. 3.8 Double conjugate points: a simple mathematical model.

rotation is resisted by the moment of inertia of the sprung mass about the same centre. The moment of inertia about $G = MK^2$. The moment of inertia about $J = I_J = M(K^2 + s^2)$. The restoring couple is supplied by the springs at front and rear. The front springs operate on an arm, $a_f = a - x + s$. The rear springs have an effective arm, $a_r = s - b - x$.

The periodic time can be calculated from these two quantities.

$$T_J = 2\pi\sqrt{\left(\frac{\text{Moment of inertia about J}}{\text{Restoring couple about J for unit angular deflection}}\right)}$$

(3.11)

$$= 2\pi\sqrt{\left\{\frac{M(K^2 + s^2)}{2k_1(a - x + s)^2 + 2k_2(s - b - x)^2}\right\}}.$$

(3.12)

Similarly, the periodic time about H is given by:

$$T_H = 2\pi\sqrt{\left\{\frac{M(K^2 + r^2)}{2k_1(a - x - r)^2 + 2k_2(b + x + r)^2}\right\}}.$$

(3.13)

Example 51

Periodic time in pitch

$$T_p = 2\pi \sqrt{\left(\frac{\text{Moment of inertia about G}}{\text{Restoring couple about G for unit angular deflection}}\right)}$$

(3.14)

$$= 2\pi \sqrt{\left\{\frac{MK^2}{2k_1(a-x)^2 + 2k_2(b+x)^2}\right\}}.$$

(3.15)

The corresponding frequencies are the reciprocals of these times.

3.4 Example

3.4.1 Ford Fiesta S

This front wheel drive saloon is an excellent example of the small popular European family car. The ride is very good for such a small vehicle. The suspension is first analysed for the case where no passengers are carried:

Data: one-up

Total mass M_t = 800 kg
Sprung mass M_s = 727 kg
Wheelbase W = 2.286 m
Front/rear weight distribution = 63/37
Front suspension rate k_1 = 21.7 kN/m
Rear suspension rate k_2 = 25.0 kN/m
$\dfrac{K^2}{(l_1 \times l_2)}$ (estimated)   = 1.0
$a = \dfrac{25}{(21.7 + 25)} \times 2.286$ = 1.224 m
$b = \dfrac{21.7}{(21.7 + 25)} \times 2.286$ = 1.062 m
$l_1 = a - x = 0.37 \times 2.286$ = 0.846 m
$l_2 = b + x = 0.63 \times 2.286$ = 1.440 m
$K^2 = 1.0 \times 0.846 \times 1.44$ = 1.218
$x = a - l_1$ or $l_2 - b$ = 0.378 m
$c^2 = a \times b$ = 1.300.

From this information, using Equations (3.6), (3.10), (3.12), (3.13) and (3.15), we obtain the following:

Double conjugate point distance r = 0.846 m
Double conjugate point distance s = 1.440 m
Front suspension frequency in bounce F_f = 1.55 Hz
Rear suspension frequency in bounce F_r = 2.17 Hz
Pitching frequency F_p = 1.96 Hz.

It is noted that the Fiesta suspension engineer has chosen double conjugate points that coincide exactly with the wheelbase, i.e.

$$r = l_1.$$
$$s = l_2.$$

This, of course, only applies to the one-up condition. As discussed earlier, when the double conjugate points are placed exactly at the wheel centres, front end bounce produces no bounce oscillations at the rear and vice versa.

Data: four-up

Total mass M_t = 1040 kg
Sprung mass M_s =   945 kg
Wheelbase W = 2.286 m
Front/rear weight distribution = 54/46
Front suspension k_1 = 21.7 kN/m
Rear suspension rate k_2 = 25.0 kN/m
$\dfrac{K^2}{(l_1 \times l_2)}$ (estimated)   = 1.05

a = 1.224 m
b = 1.062 m
l_1 = $a - x$ = 0.46 × 2.286   = 1.052 m
l_2 = $b + x$ = 0.54 × 2.286   = 1.234 m
K^2 = 1.05 × 1.052 × 1.234   = 1.363
x = $a - l_1$ or $l_2 + b$ = 0.172 m
$c^2 = a \times b$ = 1.3.

From this data we obtain the following:

Double conjugate point distance r = 1.269 m
Double conjugate point distance s = 1.074 m

Example 53

Front bounce frequency F_f	= 1.45 Hz
Rear bounce frequency F_r	= 1.69 Hz
Pitching frequency F_p	= 1.56 Hz.

This shows a general softening of the suspension.

For a small car a rear bounce frequency of about 1.7 Hz is very satisfactory. The only criticism we can make is that the rear end frequency in bounce is very close to that in pitch. The student could attempt a few experimental designs to see if he can improve on the design of the Ford Motor Co. He must always remember to work inside the physical limitations imposed by the overall body design. For example, the use of lower spring rates increases the total wheel travel. The intrusion of the upper spring mountings into the engine bay at the front, and the rear seating at the rear, could create acrimony between the suspension engineer and the body engineer.

Reference

[1] Guest, J. J. (1925–6), 'The main free vibrations of an autocar', *Proc. Inst. Auto. Engrs* (London), vol. 20, no. 505.

4
Suspension Geometry

4.1 Front wheel orientation

Over the years automobile engineers have theorized and experimented with camber angles, castor angles, kingpin inclination and offset and front wheel toe-in and still there is no consensus of opinion or general rules that can guide the student. We can only examine the steering layout and suspension geometry of such modern cars that are known to handle well and hope to glean useful pointers. Fig. 4.1 shows the two wheel angles, camber and toe-in and the two kingpin (or swivel) angles. In the left-hand view the car is viewed from the front, in the right-hand view from the side and in the lower from above.

4.1.1 Toe-in

This is a very small angle made by each front wheel plane and the longitudinal axis of the car. In Fig. 4.1 the angle is much exaggerated for the sake of clarity. The amount of toe-in is usually measured as a difference in the distance between right and left wheel rims at front and rear, both measurements being made at hub level. A typical toe-in would be as little as 3 mm. Without toe-in the inevitable compliance (or 'free play' as a typical mechanic would call it) in the several ball joints used in the steering linkages, could easily lead to the phenomenon known as 'shimmy', where the wheels flutter in and out within the limits set by the total compliance. The action of toe-in is simply to keep all the steering linkages under tension. Toe-out would also take up the slack and give stable straight-line running. The track rod and other links would be under compression in this case. This is undesirable with relatively long tubular links. If the manufacturer's specified toe-in is exceeded, excessive tyre wear will occur. The optional toe-in is usually established experimentally.

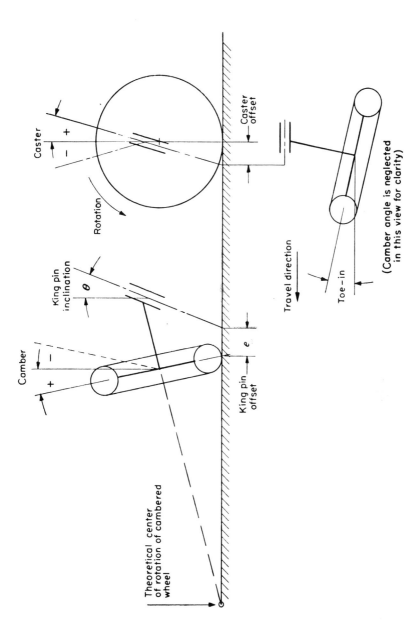

Fig. 4.1 Wheel orientation with the ground.

4.1.2 Camber

The camber angle is the angle made by the plane of the
wheel to the perpendicular as viewed from the front. It is
usually termed *positive camber* when the top of the wheel leans
outwards away from the car centre line. With modern car
tyres, camber angles cannot be large if an adequate footprint
area is to be maintained. A small amount of positive camber
at the front is traditional, since with no camber at the rear,
using the traditional live rear axle, positive camber reduces

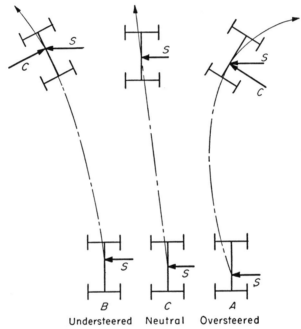

Understeered   Neutral   Oversteered

Fig. 4.2 Effect of slip angles on directional stability. Vehicles A, B
and C are subjected to identical side forces. The right-hand car A
has greater slip angles at the rear than the front. This creates a
centrifugal force C acting in the same direction (approximately) as
the side force F. This calls for rapid steering correction by the driver.
The left-hand car C has greater slip angles at the front. In this case
the centrifugal force opposes the initial side force. This is a stable
situation. Neutral steer, as shown by car B, is given when slip
angles at front and rear are identical. In this case the car suffers a
sideways displacement, but does not steer to right or left.

cornering power at the front relative to the rear. A car with greater slip angles at the front than at the rear under the action of a transient side force will be an *understeering* car. A small degree of understeer is essential for good straight-line stability. The reverse of understeer is called *oversteer* (see Fig. 4.2). This is discussed in more detail in Chapter 6.

Camber angle can change with suspension movement and this is a critical design parameter when we come to lay out the geometry of the suspension linkages. Camber angle is usually specified by the manufacturer in the unladen static condition. Static angles seldom exceed two degrees.

4.1.3 Kingpin inclination (or swivel angle)

The term *kingpin* survives from the days before independent suspension when the stub axle on which the wheel rotated

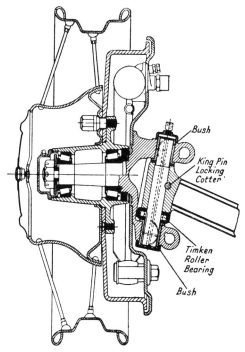

Fig. 4.3 Details of kingpin and stub axle on beam front axle.

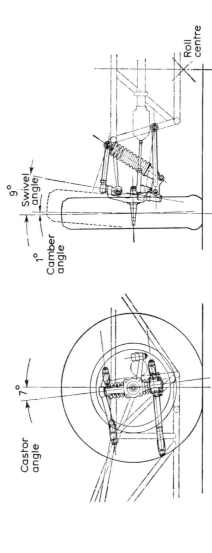

Fig. 4.4 Modern steering layout, using upper and lower swivelling joints to replace kingpin.

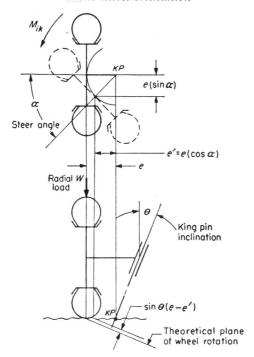

Fig. 4.5 Effect of kingpin inclination on steering. When wheel is steered around inclined kingpin, the axle is lifted, creating an unstable condition. This provides a part of the self-aligning torque.

was arranged to swivel around a hardened steel pin mounted in the end of the beam axle (see Fig. 4.3). With the introduction of independent front suspension the kingpin survived for a time, but was eventually replaced by a system in which the hub carrier swivels about an upper and lower ball joint, as shown in Fig. 4.4.

The kingpin or swivel angle is chosen to give the desired offset, as shown in Fig. 4.1. The extent of this offset determines the amount of self-aligning torque exerted when the steering wheel is turned. This is illustrated in Fig. 4.5. Turning the wheel, raises the wheel hub relative to the ground. A moment is thus created which tends to return the wheel to the straight-ahead position. Some modern cars have been given negative offset. This reduces steering 'feel' at the steering wheel, but is introduced to give stability at speed in the event of a tyre blowout or brake failure on one front wheel.

4.1.4 Castor angle

Castor angle also creates self-aligning torque, since it places the contact patch behind the swivelling axis (see Fig. 4.1). Cars with a kingpin angle that gives a negative offset will, therefore, need a larger castor angle to compensate.

4.1.5 The dynamics of self-aligning torque

Self-aligning torque defies the rules of simple geometry, since the behaviour of the tyre footprint exerts a major influence as the slip angle rises. For moderate values of cornering force self-aligning torque increase with slip angle. At extreme slip angles the self-aligning torque is reduced, as shown in Fig. 4.6. Only when cornering at racing speeds is this reversal encountered. To a racing driver, this loss of 'feel' at the steering wheel is an indication that he is now driving 'on the limit'.

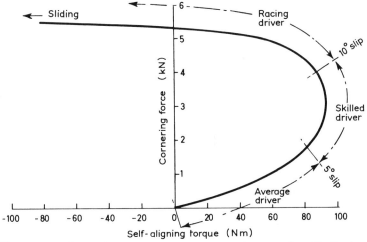

Fig. 4.6 Relationship between cornering force and self-aligning torque on a typical tyre.

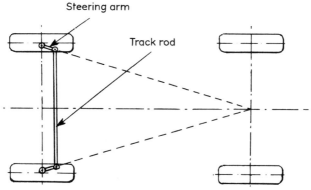

Fig. 4.7 Ackermann steering layout.

4.1.6 Ackermann steering layout

Ackermann arranged his steering layout to give a slightly greater steering angle to the inner wheel than the outer. His original layout is shown in Fig. 4.7. The steering arms are inclined inwards so that their projected lines converge to meet at the centre of the rear axle. If one neglects the effects of slip

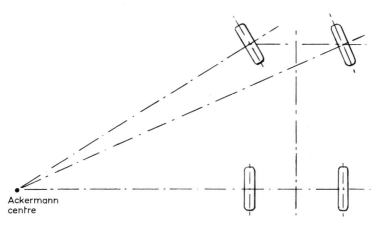

Fig. 4.8 Ackermann steering centre (with zero slip angles).

angle (as Ackermann obviously did), the car will have a turning centre on any corner which is in line with the rear axle centre line (see Fig. 4.8). When slip occurs at the four contact patches the turning centre will move forward, the extent of this movement depending upon the relative slip angles at front and rear. Not all designers adopt the original Ackermann layout. Some use an intersection point behind the rear axle, others prefer parallel steering arms. There is some logic in adopting a negative Ackermann angle, i.e. one where the intersection point lies ahead of the front wheels. High cornering forces create roll and demand greater slip angles from the outer tyres than the inner. The original Ackermann layout, designed to give a greater angle to the inner wheel, is therefore based on a false premise. The dynamic behaviour of a car when cornering will be analysed in Chapter 6.

4.2 Roll geometry

4.2.1 Roll resistance

When a body rolls under centrifugal force, this force acts at the centroid of the sprung mass (see Fig. 4.9). The body rotates about the *roll centre* and this most important parameter

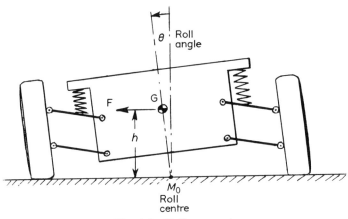

Fig. 4.9 Roll moment.

is defined by the geometry of the suspension system, as will be discussed later in this chapter. The roll moment which must be resisted by the suspension springs (and the anti-roll bars, when these are fitted) is equal to $F \times h$, where F is the centrifugal force and h is the distance between the sprung mass centroid and the roll centre. For a given spring rate, the roll resistance will be greater and the roll angle less if h is reduced.

4.2.2 Roll centres

For a typical double wishbone suspension the position of the roll centre can be determined as follows. For simplicity, only one end of the car is considered at this stage. Alternatively, we could say that the front and rear suspensions are considered to be identical, the weight distribution 50/50 and the axial centroid centre line horizontal. Under a side force F, the body will tilt and the spring on the left will be compressed to increase the vertical load R_1 (see Fig. 4.10). We could produce the same effect by holding the body in a horizontal position while rotating the wheel L about the *instantaneous suspension centre* C_L relative to the body. This centre C_L is given by the intersection of the projected centre lines of the upper and lower wishbones. The wheel footprint

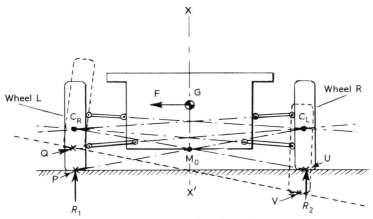

Fig. 4.10 Determining the roll centre.

will move at right angles to line PC_L to a new position Q. At the same time the decrease in load R_2 will cause the contact point U of wheel R to move to position V. If we rotate the body in an anti-clockwise direction until the wheel footprints have returned to their original positions P and U, the only point about which rotation can occur within the constraints that wheel L must move about centre C_L and wheel R move about centre C_R is seen to be point M_0. The position of M_0 is found by the intersection of lines drawn between C_L and P and C_R and U. In some cases, notably the popular MacPherson strut suspension, the roll centre is not at a constant height above ground and moves slightly as the roll increases.

4.2.3   Typical roll centres

The roll centre on a beam axle supported on semi-elliptic springs is decided largely by the curvature of the springs and the position of the shackles. In general it is situated approximately halfway between the spring base and the upper shackle pin, as shown in Fig. 4.11.

Three systems have the roll centre at ground level. The first is a double-wishbone system with parallel links of equal

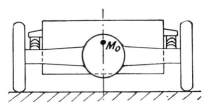

Fig. 4.11   Roll centre on beam axle with semi-elliptic springs.

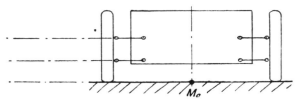

Fig. 4.12   Roll centre with double-wishbone suspension of equal-length parallel links.

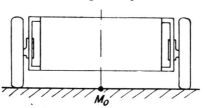

Fig. 4.13 Roll centre with vertical sliding pillars.

length (see Fig. 4.12). Since the instantaneous link centres are at infinity, the roll centre must be at ground level. The second system is the old Porsche trailing link system using equal-length parallel links as used on the familiar VW Beetle. This system is shown in Fig. 5.10 in the next chapter. The third is the sliding pillar suspension with vertical pillars, as shown schematically in Fig. 4.13. When the sliding pillar is set at a swivel angle (see Fig. 5.9 in the next chapter), the roll centre falls below ground centre. The position is defined by extending the pillar centre line to intersect the ground plane, then by drawing a perpendicular line to intersect the vertical centre line of the car.

The de Dion suspension system separates the two functions normally performed by a live rear axle. The final drive unit is mounted on the chassis (or body) and is, therefore, part of the sprung mass. Open halfshafts with universal joints at both ends take the drive to the wheels. The de Dion tube which connects the two wheels and maintains them in perfect parallelism is always located laterally, either by a vertical slide, as shown in Fig. 4.14, or by other effective means. With effective lateral location, the body is constrained to roll about the centre point of the de Dion tube.

The swing axle is now obsolescent. It was used by Professor Porsche for the rear suspension on many of his

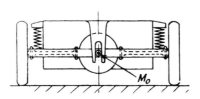

Fig. 4.14 Roll centre with de Dion system.

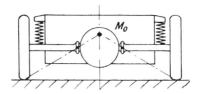

Fig. 4.15   Roll centre with swing axle suspension.

designs, almost to the point of obsession. The system became notorious for excessive roll-steer effects that militated against safe handling. Fig. 4.15 shows that the roll centre is very high. This, of course, tends to reduce the roll moment (see Fig. 4.9) and it is theoretically possible to design a racing car in which the centroid of the sprung mass coincides with the roll centre. This, at first sight, is an ideal system. Unfortunately the cornering force, acting at road level on the footprint of the inner wheel, tends to twist the swing axle and lift the inner end. This produces what is called a 'jacking action'. Fig. 4.15 shows how this can happen at high cornering forces. The Daimler-Benz Co. eventually produced a low-pivot swing axle design. It was a complicated, but very effective, design (see Fig. 4.17) and was eventually replaced by the semi-trailing link design which is so popular today and which will be described in Chapter 5.

4.2.4   MacPherson strut

This type of suspension – sometimes called a Chapman strut – is well illustrated in Fig. 4.18 and the schematic drawing in Fig. 4.19 shows how the roll centre is found. The hub carrier slides up and down the strut but is attached rigidly to the outer tubular member. The strut must be capable of a slight angular movement at its upper attachment point, since the lower link moves through an angle under bump and rebound. Some compliance is, therefore, necessary in the upper attachment. This is usually given by the provision of a rubber mounting. Some designers also provide a ball thrust bearing, others use a low-friction thrust washer. This permits strut rotation as the wheels are steered.

Fig. 4.16 Jacking action on swing axle suspension (*Motor* photograph).

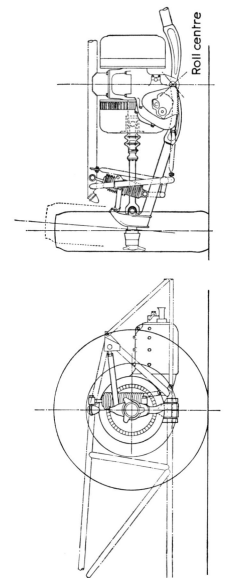

Fig. 4.17 Low pivot swing axle suspension, as used on Mercedes-Benz W196 racing car.

Roll centre

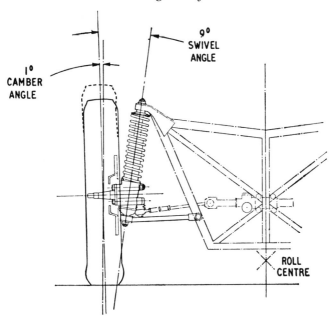

Fig. 4.18 Chapman strut suspension on early Lotus sports car.

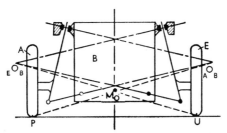

Fig. 4.19 Determining centre on MacPherson (Chapman) strut suspension.

4.2.5 Roll axis

The roll centres at the front and rear are seldom at the same height. Similarly, the static loads carried at the front and rear are not often equal. Fig. 4.20 shows how the sprung mass can be treated as two discrete masses, M_f at the front and M_r at the rear. The centroid axis is at an angle relative to the roll

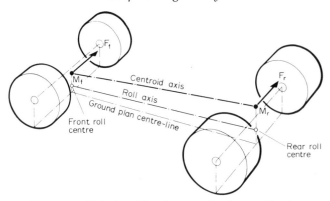

Fig. 4.20 Relationship of centroid axis to roll axis.

axis, this angle being determined by the heights of the front
and rear masses M_f and M_r above their respective roll centres.
In the particular example the roll moment is obviously greater
at the rear. The suspension system must be designed to cope
with this imbalance.

4.3   Anti-roll bars

Anti-roll bars often supply a simple solution to the problem
of roll imbalance. It is a simple device, being a transverse bar,
cranked at both ends in the same direction and clamped to
some convenient point on one of the suspension links on
opposite sides. They can be fitted to front or rear suspensions
or to both.

It will be seen in Fig. 4.21 that body roll will cause one
cranked arm to move downwards, while the other moves upwards. The anti-roll bar is, therefore, a simple torsion bar. If
however both wheels at the same end of the car rise or fall in
unison, no torsion is applied to the bar. Under the action of
single wheel bounce, the main spring on this side works in
parallel with an effective additional spring which is the
combination of the anti-roll bar and the main spring on the
opposite side working *in series*. To clarify what this statement
means, let us first consider what happens when two springs
of different rates act together *side by side*, i.e. in parallel. For

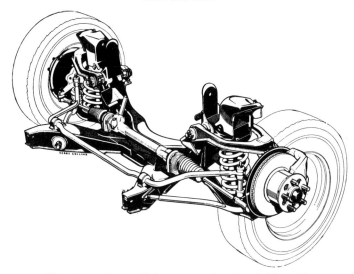

Fig. 4.21 Anti-roll bar on Ford Granada/Consul.

springs in parallel the rates are additive:

$$k_p = k_1 + k_2 \qquad (4.1)$$

where k_p is a single spring of equivalent rate to the two springs of rates k_1 and k_2 acting in parallel.

When two springs act *in line*, i.e. in series, the equivalent spring is always of reduced rate. An equivalent spring of stiffness k_s to two springs k_1 and k_2 acting in series must produce the same extension under the same load:

$$\frac{M}{k_s} = M\left(\frac{1}{k_1} + \frac{1}{k_2}\right)$$

$$= \frac{M}{(k_1 k_2)/(k_1 + k_2)}.$$

Therefore

$$k_s = \frac{k_1 k_2}{k_1 + k_2}. \qquad (4.2)$$

4.3.1 Single wheel bounce (with anti-roll bar)

Applying the above to the case of single wheel bounce, we can take the main spring rate at the wheel as k and the anti-roll bar rate, also measured at the wheel, as k_a. The single wheel bounce rate k_e is, therefore, given by:

$$k_e = k + \frac{k_a k}{k_a + k} . \qquad (4.3)$$

A good example to illustrate how the anti-roll bar can influence the single wheel bounce rate is the Porsche 928, which is the subject of a design analysis in Chapter 12. The Porsche 928 is provided with a strong anti-roll bar at the front and a relatively weak one at the rear. The appropriate data are as follows:

	Front	*Rear*
Wheel rate (kN/m)	18.63	22.55
Anti-roll bar rate (at wheel)(kN/m)	83.4	10.2.

Single wheel rates in bounce

$$Front = 18.63 + \frac{83.4 \times 18.63}{83.4 + 18.63}$$

$$= 33.86 \text{ kN/m}$$

$$Rear = 22.55 + \frac{10.2 \times 22.55}{10.2 + 22.55}$$

$$= 29.57 \text{ kN/m.}$$

This demonstrates the penalty involved when relatively stiff anti-roll bars are resorted to as a means of limiting roll angles.

4.4 Anti-dive and anti-squat

Without resort to anti-dive and anti-squat geometry in the suspension layout, cars with soft suspension will dip at the front and rise at the rear under heavy braking. During hard acceleration the reverse effect will occur. It was not uncommon more than twenty years ago to see American

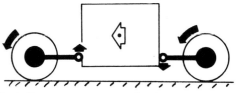

Fig. 4.22 Anti-dive system, using leading arms at the front and trailing arms at the rear.

sedans perform something that resembled a curtsy when stopping at the traffic lights. A source of embarrassment was the locking of front bumpers under the rear bumpers of the car immediately ahead. It was no easy task to separate the locked cars. The provision of over-riders on bumpers prevented this particular danger but the physical discomfort still remained. Brake dive is felt by the passenger as a pitching motion and this puts more strain on the neck muscles than is experienced with simple deceleration in the horizontal plane.

A simple anti-dive design could be given by the use of a leading arm at the front and a trailing arm at the rear, as shown in Fig. 4.22. Under braking action the backplates (or callipers, when disk brakes are fitted) tend to rotate with the wheels. This produces an upward reaction at the front of

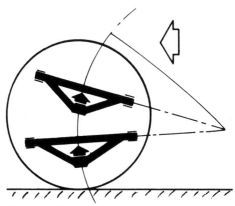

Fig. 4.23 Angled double-wishbones to give effective leading arms to front suspension.

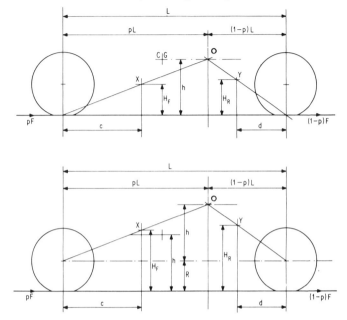

Fig. 4.24 Anti-brake dive geometry: (a) with outboard brakes; (b) with inboard brakes.

the body and a downward at the rear. With the correct proportions, this can be designed to give full anti-dive reaction.

Fig. 4.23 shows how the upper and lower links of a double-wishbone front suspension system can be angled to give an effective leading arm length. Fig. 4.24 applies to such a case. It is assumed that the front braking effort does not vary from the designed front/rear bias. If the front/rear braking effort is designed to give an effort ratio of 60/40, the value of p in both upper and lower diagrams of Fig. 4.24 will be 0.6.

The upper diagram applies to the case of a car with outboard brakes. Lines are drawn from the front and rear footprints to point O. This point is on the sprung mass centroid axis, the position relative to the front and rear wheels being determined by the braking ratio $p/(1-p)$. For full anti-dive correction, the effective pivot points, X for the front suspension and Y for the rear suspension, should lie on the appropriate lines. With inboard brakes, since the brakes are part of the sprung mass, the braking torque is applied through the drive shafts. The suspension pivots in this case must therefore lie on lines drawn between the wheel centres and point O, as shown in the lower diagram.

Anti-dive geometry is not without its problems. Since the upper and lower wishbones pivot about non-parallel axes, this results in a change in the front castor angle with bump and rebound. Some designers find this castor angle variation unacceptable and compromise by correcting only a percentage of the brake dive. The Jaguar XJ-S, for example, is designed to give only 50% anti-dive correction.

4.4.1 Anti-squat

Anti-squat is resistance to squatting down at the rear end, and as we may see in its more spectacular forms on the drag strips it can be resisted in the limit by the provision of rollers at the rear extremities of the body. Fortunately for the designer of more ordinary vehicles, the provision of 'leading arm effect' at the front and 'trailing arm effect' at the rear works well in the reverse direction. Some compromise is necessary, since an anti-dive geometry carefully calculated to match a particular front/rear brake distribution will seldom give a perfect correction for anti-squat.

In Formula 1 racing the problems of anti-dive and anti-squat became acute in 1979, when the majority of Formula 1 cars adopted 'ground-effect' body designs that pulled the car down on the road by the vacuum created on the underside of large side pods. Sliding skirts were devised to seal the body sides to prevent leakage of higher-pressure air to destroy this vacuum. These skirts actually touched the road surface and often made grooves in the tarmacadam when the suspension exceeded the available movement of the skirt in its slides.

The most successful designs of the 1979 racing season were those that achieved the closest approach to a no-roll suspension with negligible dive and squat. The latest type T4 version of the Ferrari 312 uses heavy Bowden cables running from front to rear to give a coupled front/rear suspension system designed to reduce dive and squat. As I write, the outcome of the experiment is unresolved, but at least they have won the Constructors' Championship for the year.

5
Conventional Systems

5.1 The beam axle

Less than fifty years ago the beam axle suspended on semi-elliptic springs was the norm at both ends of the automobile. There were a few exceptions. Sizaire-Naudin used a sliding pillar independent suspension at the front as early as 1906 and a few years later the Morgan three-wheeler and the Lancia Lambda appeared with refined examples of the basic design. Even so, these were regarded as rather expensive eccentricities and did little to convince the bulk of automobile manufacturers that the simple rigid front axle mounted on leaf springs, and the much heavier rear axle casing with similar springs, were not a final solution to the problem of springing. All that was needed, or so it seemed, was a measure of refinement and a gradual reduction in the manufacturing costs.

Despite this air of self-satisfaction, many experienced motorists knew that both front and rear axles could sometimes develop their own form of 'shakes'. At the front it was called *shimmy*, at the rear *tramp*. For good measure, some front axles were capable of both phenomena.

5.1.1 Shimmy

The driver could feel the onset of shimmy when the steering wheel began to vibrate rapidly from side to side. This, precisely, was what was happening to the front wheels. Accessory manufacturers made steering dampers designed to reduce the amount of shimmy. In racing and sports cars the designer provided very stiff springs to reduce the angular deflection of the front axle, since this angular deflection of the axle is the source of shimmy.

Any rotating mass, such as a wheel and tyre, exhibits gyroscopic effects. Thus, if a spinning wheel is subjected to an angular displacement, a force is produced tending to move the axis of rotation in a direction at right angles to the original displacement. The reader can demonstrate this phenomenon using an ordinary bicycle, preferably one with large-diameter wheels. He should stand astride the bicycle and lift the front wheel clear of the ground. Then holding the cycle in this position with one hand only on the handlebars, spin the front wheel as fast as possible in the normal forward direction. Finally, with the wheel spinning fast, he should grasp the handlebars with both hands and lean the cycle to the right. The gyroscopic torque will be felt as a distinct reaction at the handlebars as the wheel tries to turn to the right. If the wheel is spun in the reverse direction, the steering reaction will be to the left.

This serves to illustrate the gyroscopic torque that twitches the front wheels to left or right when passing over a bump or a pothole in the road (see Fig. 5.1). With inadequate damping of the front suspension (a characteristic of early vehicles), the inertia of the heavy front axle and the masses of the wheels at each end induces an oscillation at the natural frequency of the springs. As long as these oscillations continue, with one wheel rising as the other falls, the gyroscopic torque will produce the steering oscillations of shimmy.

5.1.2 Tramp

With a beam axle tramp can occur at front or rear. Even when shimmy is eliminated a beam axle can still develop tramp, which is a transverse rocking action at spring frequency that can persist for many seconds. The most common cause of

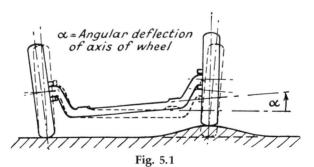

α = Angular deflection of axis of wheel

Fig. 5.1

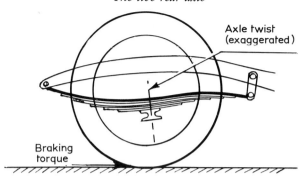

Fig. 5.2 Axle twist under braking torque.

front axle tramp is induced by hard braking. Even the great Henry Royce was shaken physically and mentally when he tested the first Rolls-Royce to be fitted with front wheel brakes. Brake tramp was so violent that the headlamp bulbs failed as the filaments broke. For a time Henry Royce was in the dark! Improved location of the front axle was the eventual solution.

The conventional location of semi-elliptic springs at this time was through the single shackle pin at the front. Variation in spring length was accommodated by the provision of swinging shackles at the rear. Under braking torque the springs could 'windup', i.e. twist into a flattened S-shape as shown in Fig. 5.2. Spring windup reduced the castor angle. In extreme cases it could become negative, resulting in a negative self-aligning torque.

I remember a bad case of brake tramp when a vintage car was raced for the first time at a VSCC Silverstone meeting. Hard braking causes such severe brake tramp that the magnificent mahogany dashboard was cracked from side to side.

Some designers of vintage sports cars used torque rods to prevent spring windup. Others found an easy solution in the adoption of very stiff springs that gave a very hard ride.

5.2 The live rear axle

All automobile designers have now adopted independent suspension for the front wheels. A beam axle at the rear is usually called a *live rear axle* when it also acts as the housing for the final drive unit and axle shafts. The term *dead rear axle* has been adopted to describe the use of a beam axle between the rear wheels on a front wheel drive car.

5.2.1 Disadvantages

The live rear axle is relatively heavy in comparison with an independently suspended one. The outer casing is constructed from malleable castings and/or steel pressings and contains the final drive crown wheel and pinion, the differential casing and its gears, the halfshafts and all the necessary bearings to support the moving parts. The customary drum brakes, brake shoes, backplate, wheel hubs, wheels and tyres all contribute to a formidable unsprung mass.

The unsprung mass is approximately twice that of a good design of independent rear suspension. This in itself is a serious disadvantage as was demonstrated in Chapter 2. A car with a low ratio of unsprung-to-sprung mass (m/M) will maintain tyre/road contact on a rough road much more effectively than one with a high ratio. In Chapter 2 it was also shown that a high m/M ratio is also associated with a hard ride.

Apart from the low cost of the live rear axle, it does represent a very robust design and with the help of modern techniques of longitudinal and transverse location satisfactory rear suspension can be given for a low-cost front engine/rear drive family car. The word 'satisfactory' is used advisedly, since the high standards established by a good design of IRS cannot be matched.

5.2.2 Five-link system

Paradoxically this popular location system for the live rear axle owes much to the experience gained in the development of IRS systems intended as replacements for the live rear axle. The five-link system uses two non-parallel unequal-length trailing links per side to give longitudinal location of the axle (see Fig. 5.3). Parallel links of equal length would give the closest approach to vertical motion of the axle under bounce

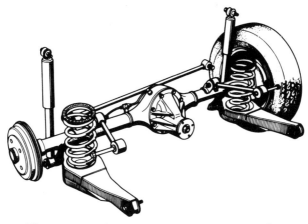

Fig. 5.3   Five-link suspension on live rear axle.

and rebound, but the links are usually given a convergence on an effective pivot centre, thus giving a measure of anti-dive effect. The fifth link is required to give lateral location. A Panhard rod, as shown in Fig. 5.3, does not give perfect location but is a simple solution. The rod should be as long as possible to constrain the axle to move in a 'flat arc'. One end of the rod is attached to a bracket on the body/frame, the other to a bracket on the axle casing on the opposite side. Rubber bushes are used on all five links to reduce the transmission of vibrations to the body.

There are alternative systems of lateral location. A popular one giving perfect lateral location over a limited range, is that

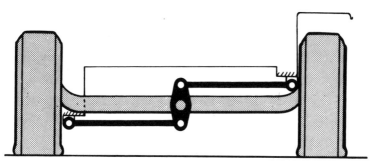

Fig. 5.4   Watt's linkage location for de Dion suspension.

invented by James Watt. Fig. 5.4 is a simple diagram to explain the principle of this linkage. With such excellent axle location, one can replace the heavy semi-elliptic spring with the lighter, less expensive, coil spring. This again is a technical spinoff from the development work on independent suspensions.

5.2.3 Four-link system

This system dispenses with the lateral location link by arranging the 'longitudinal' upper links at an angle relative to the centre line of the car. Such a layout is not likely to give such accurate lateral location as that provided by a Watt's linkage.

5.2.4 Two-link plus A-bracket system

This is a variation on the above theme. There are two lower links (as in the above systems), but the upper links are integrated into a single A-bracket pivoted at two points on the body and one above the centre point of the axle. The Renault, shown in Fig. 5.5, is a good example of this system applied to a dead rear axle.

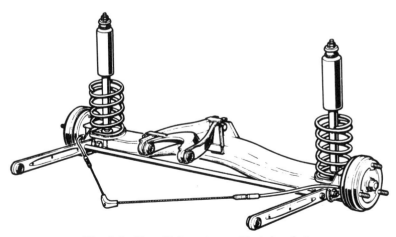

Fig. 5.5 Two-link system with A-bracket.

5.2.5 Torque tube

Since the early designs of an open propeller shaft were so unreliable, it was common to enclose the drive in a casing or torque tube and let this and the whole rear axle casing pivot about a spherical bearing at the rear end of the gearbox. The complete T-shaped assembly was thus given an accurate location, even though the unsprung mass was increased considerably by the long torque tube.

Hotchkiss was a pioneer in the use of an open propeller shaft, using the semi-elliptic springs to locate the rear axle. Surprisingly he also pioneered the use of a short torque tube in conjunction with a short open propeller shaft in the 1922 type AR Hotchkiss. In this large luxury car a cruciform cross-member is provided approximately halfway between the gearbox and the rear axle. The spherical joint of the torque tube is mounted at this station. This not only reduces the mass of the torque tube, but removes the whirling dangers associated with the use of very long open propeller shafts. There is, therefore, much to recommend the layout

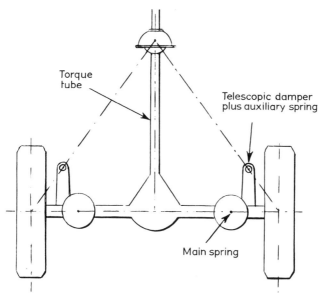

Fig. 5.6 The Rover 3500 live rear axle with short torque tube.

chosen for the Rover 3500 as shown in Fig. 5.6. The coil springs are placed immediately above the axle tubes. Forward of each tube are combination spring/damper units. These are located on forward-facing brackets at points on the lines drawn between the torque tube universal joint and the wheel centres. In this position the spring/damper units can control body roll as well as bump and rebound. Lateral location is given by a Watt's linkage.

5.3 The de Dion axle

The invention of the de Dion axle is usually attributed to Comte Albert de Dion, but it was actually conceived by two engineers, Trépardoux and Bouton, who worked for him. A well-engineered design that serves to illustrate the principle is that used on the 1938 type W154 Mercedes-Benz Grand Prix car (see Fig. 5.7). The wheel hub carriers are coupled together by a cranked de Dion tube, which in this design can pivot at its centre point and is constrained to rise and fall in a vertical slide. The sliding block is pivoted at the front of the tube, while the slide is incorporated in the final drive casting. Unlike a live rear axle, the final drive unit is carried on the chassis and is thus part of the sprung mass. The drive to the wheels is by universally jointed halfshafts. A single ball-jointed radius arm is attached to the top of each wheel hub carrier. To permit single wheel bump and rebound, provision was made for relative movement between the right- and left-hand halves of the de Dion tube. The provision of twin parallel, equal-length trailing links would have removed the need for this complication (see Fig. 5.8).

The de Dion axle retains the one advantage of the beam axle, in that the wheels are maintained parallel at all times. When cornering on a good surface, both wheels remain vertical. This was advantageous in 1938. In the 1980s, with more and more cars using low-profile tyres, it is still an excellent feature. The other advantage is that the unsprung mass is much reduced in comparison with that of a live rear axle. As will be shown later, it is difficult to achieve as low an m/M ratio as that given by the lightest designs of independent rear suspension.

Fig. 5.7   The 1938 Mercedes-Benz type W154 de Dion axle.

Aston Martin have used a de Dion rear axle on their cars for years with excellent results. As will be seen from Fig. 5.8, the unsprung weight has been kept to a minimum by the use of inboard disk brakes. The provision of parallel trailing arms on each side also makes it possible to use a one-piece de Dion tube. Transverse location is given by a Watt's linkage.

The drive shafts are fitted with universal joints at both ends. Either the inner or the outer joints must permit some axial movement. On the Aston Martin the variation in drive shaft length under bump and rebound is accommodated by the provision of sliding splines inside the inner universal joints.

5.4   Independent front suspension

5.4.1   Sliding pillar system

We have already said enough about the disadvantages of a live front axle to show how necessary it was to replace it. Sliding pillar IFS was first used in 1906 and, if we neglect the rather primitive designs that appeared on early Morgan three-wheelers, it appears that the first successful design appeared on the Lancia Lambda, first shown at the Paris

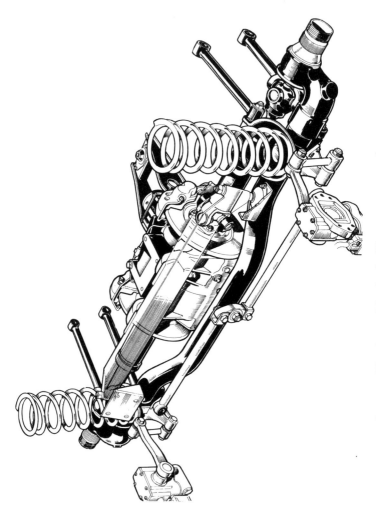

Fig. 5.8 The Aston Martin de Dion rear suspension.

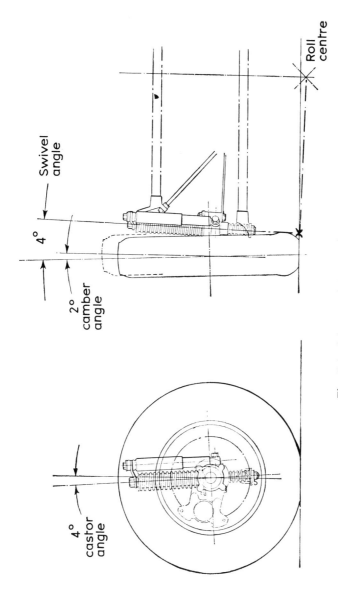

Fig. 5.9 Morgan sliding pillar IFS.

Salon in 1922. Even by modern standards, the road-holding of this vintage treasure is very good. On the Lambda each wheel hub carrier moves up and down on a sliding pillar. Since the pillars are integral with a transverse tubular frame that is part of the chassis, the unsprung weight given by this system is probably the lightest ever devised. The Lancia system was very sophisticated. The coil springs were enclosed and tubular hydraulic dampers were incorporated. (Unfortunately, I am not able to find a suitable illustration of this excellent example.)

The only surviving example of this sliding pillar system is seen on the Morgan sports car. From the earliest designs dating back to 1910, the Morgan Motor Co. has gradually developed the well-engineered layout shown in Fig. 5.9. Single wheel movements in bump or rebound are vertical, with a small reduction in track, as shown by the broken line in the figure as the wheel rises. When the car rolls, however, both wheels adopt new camber angles directly related to the roll angle. In the example shown with static camber angles of 2°, a role of 3° when cornering hard would give a camber of 5° on the outer wheel and 1° on the inner. The reverse would be more acceptable with modern low-profile tyres. To combat this disadvantage, relatively stiff springing is necessary on the Morgan.

5.4.2 Trailing link suspension

Professor Ferdinand Porsche made this system memorable when he adopted it for the Volkswagen Beetle. Fig. 5.10 shows an early design of Porsche front suspension as used on the type 356B Porsche. With the tyres of the period (*circa* 1960) with aspect ratios no less than 85% the contact patch was well maintained and suffered little loss of area up to roll angles of 5° or 6°. This of course is only one aspect, since a contact patch becomes less effective at a positive camber angle and with a trailing link suspension both inner and outer wheels adopt camber angles equal to the roll angle. The loss of cornering force can be seen from the typical curves shown in Fig. 5.11.

As with the sliding pillar suspension, large camber changes are anathema when low-profile tyres are used. The Porsche Co. abandoned this type of suspension later in the 1960s.

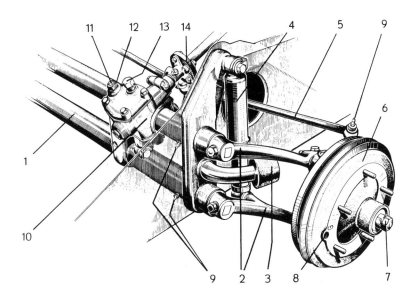

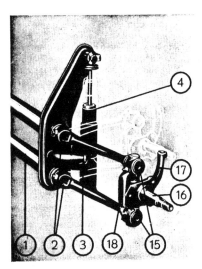

Fig. 5.10 Parallel trailing link suspension on Porsche type 356B: 1, front axle tubes; 2, torsion arm; 3, rubber buffer; 4, telescopic shock absorbers; 5, tie rod; 6, brake drum; 7, adjusting nuts for front wheel bearing; 8, opening for brake adjusting; 9, grease nipple; 10, steering gear; 11, adjusting screw for sector shaft; 12, hexagon nut for adjusting screw; 13, screw plug for oil-filling port; 14, coupling disk; 15, torsion arm link pin; 16, stub axle (steering knuckle); 17, steering arm at stub axle; 18, suspension arm link for stub axle.

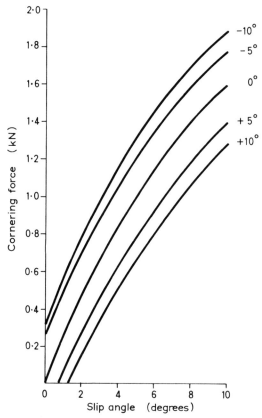

Fig. 5.11   Influence of camber angle on cornering force.

5.5   Wishbone systems

5.5.1   Parallel links

Earlier wishbone systems usually had parallel links, often of equal length as in Fig. 5.12. In this 1935 R-type MG both front and rear suspensions were by parallel double-wishbones. These wonderful little single-seater racing cars skated round corners with all four wheels at astonishing camber angles. The low m/M ratio helped to keep the tyre

Fig. 5.12 The R-type MG Midget.

footprints glued to the road surface, but the loss of cornering power from the positive camber angles (about 10% at 5° roll angle) threw away some of the inherent superiority of independent suspension.

The live beam axles used by their rivals had small front camber angles and zero camber angles at the rear. At this time, one must remember that the importance of camber angle was not fully appreciated and the techniques by which tyre makers now measure cornering force had not yet been developed.

5.5.2 Angled double-wishbones

Under this heading we can consider true double-wishbone systems or related systems such as the one popular on racing cars in which a normal wishbone is used at the top and a

Fig. 5.13 Jim Clark's Formula Junior Lotus cornering at Oulton Park. Note verticality of outer wheels.

single lower link. Sometimes the lower link is a single or double longitudinal torque rod used to resist braking forces. If we consider these three versions under one heading, they represent one of the most popular layouts in use today.

There are four design parameters we can use to make changes in the roll centre height and in the camber changes and track variation with bump and rebound. These are:

(a) the relative angle between upper and lower links;
(b) the angle the lower link makes with the horizontal;
(c) the relative lengths of the upper and lower links;
(d) the ratio of lower link length to track.

The larger car manufacturers are able to use computer studies to 'optimize' their suspension systems. As already stressed, there is no perfect solution. The suspension engineer must always choose a compromise. It is still possible, if one is prepared to burn the midnight oil, for the young designer to find a well-balanced design. Racing car designers have been doing it for years.

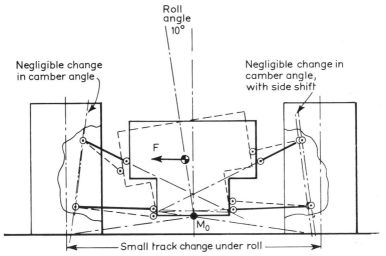

Fig. 5.14 Angled double-wishbone system for racing car.

For twenty years designers of racing cars and competition sports cars have aimed at a suspension geometry that will maintain verticality of the more heavily loaded outer wheel under all angles of roll. Jim Clark's Formula Junior Lotus is seen in Fig. 5.13 with both front and rear outer wheels very close to the vertical in the middle of a tight bend. The camber of the lightly loaded inner wheels is obviously regarded as unimportant. Fig. 5.14 has been constructed to show that one can juggle with the four parameters stated above and devise an angled double-wishbone system which will maintain both outer and inner wheels very close to the vertical, even at the extreme roll angle of $10°$. A typical racing car seldom exceeds a roll angle of $1.5°$.

5.5.3 No-roll layout

A pertinent question that must already have been asked by many readers is: Why not choose a geometry that gives a roll centre that coincides with the centroid? In this way roll would be eliminated. Why not indeed!

We can only answer this question by asking another: Do we really want to eliminate roll? It can be argued that roll to the average driver is a rough indication of the extent of the

centrifugal force, but the medical profession has certain
reservations on the accuracy of this indication, particularly at
high values of g.

5.5.4   The physiology of roll

The human body was not designed (perhaps Charles Darwin
would dispute the use of this word) to interpret turn signals
from the antics of aeroplanes or automobiles. We try to
interpret the messages we receive through the middle ear, a
balance mechanism developed over millions of years to help
us to walk upright; certainly not to drive a car. The muscles
of the thighs and buttocks can be used to measure the
intensity of the side force produced by centrifugal force when
firmly held in a well-designed seat, but this again is not a
natural reaction. Early aviators complained that the parachute
upon which they were asked to sit turned them into 'deadend
kids', since this reduced their ability to sense the signals
transmitted to them by their buttocks.

Contrary to popular belief, the semi-circular canals in the
middle ear contribute very little to our sense of sideways
acceleration. The semi-circular canals are, of course, aware of
roll and many drivers react to the balance signals from these
canals by leaning away from the angle of roll. Underneath
these canals are two very important saucerlike disks in which
tiny hard particles are suspended in a jellylike substance.
These particles, called otoliths, collide with sensitive hairs
attached to nerve cells when the middle ear is subjected to
acceleration. These otolithic organs can detect very small
deviations from the vertical, as demonstrated by so many
circus balancing acts. When a racing car is cornering, these
organs measure the roll angle *plus* the lateral acceleration. It
would, therefore, be difficult for the driver to differentiate
between these two factors and estimate his limiting cornering
speed if he had to rely on the otolithic organs alone.
Fortunately, if my analysis is correct, he also has the reaction
felt through the muscles of his thighs and buttocks. No
experimental evidence exists to support this hypothesis, but it
is interesting that all racing drivers insist on the necessity of a
well-tailored seat and their attendance at the works for a
fitting of the moulded seat contours is as much a ceremony as

Bertie Wooster's visit to his Savile Row tailor. With cornering accelerations now reaching values of 2*g*, the strain on the gluteous muscles of the upper thighs, on the back muscles and certainly on the muscles of the neck are very high indeed, so high perhaps that fatigue could begin to cloud a driver's judgment towards the end of a race.

Over the years the roll angles on racing cars have become less and less. Today they seldom approach 1.5° when cornering at the limit. With such a small roll angle, the guidance given to the driver from his otolithic sense organs will be produced almost entirely by the centrifugal acceleration. In the case of a racing car, therefore, the driver has two senses to guide him when cornering. Modern family saloons often roll at angles ranging from about 4° to as much

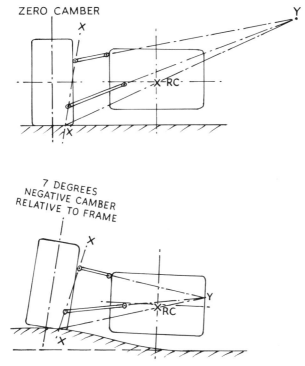

Fig. 5.15 Angled double-wishbone system with very high roll centre. Effect of single-wheel bump.

as 8° under lateral accelerations of 0.5*g*. The middle ear, with its two sensing functions, must therefore be regarded as an unreliable guide to the safe accelerations in a corner. Experience of a particular model is still the safest way to build up knowledge of the way a car handles in a corner.

No-roll suspensions have been the subject of several interesting developments, as will be described in Chapter 10. If we raise the roll centre on an angled double-wishbone system until it coincides with the centroid of the sprung mass, we now find that we have constructed a system that gives undesirable changes in camber angle under single wheel bump and rebound. This is illustrated in Fig. 5.15, where both upper and lower wishbones are angled upwards towards the car's centre line to give a high roll centre. In this way one could construct a no-roll suspension, but the camber changes in single wheel bump and rebound would be simply unacceptable, as shown in the lower diagram.

5.5.5 MacPherson strut suspension

This system was invented by Earle MacPherson of General Motors about thirty-five years ago. A good example of the system is that fitted to the Datsun 260Z and 280Z sports cars. A cross-sectional front elevation is given in Fig. 5.16. Upper left is a plan view, giving a clear indication of the anti-roll bar (stabilizer bar) and the compression rod, which resists braking forces. Since the Datsun 260Z has rear wheel drive, the compression rod is also under compression during acceleration.

The construction of the MacPherson strut is typical. The hub carrier is attached (by welding) to the outer tubular member of the strut. At the base of this member is a sealed ball joint mounted in the extremity of the lower transverse link. Under maximum bump and rebound, the strut moves through an angle of 2.5° and the required compliance is given by a hollow rubber bush where the strut is attached to the body. This rubber bush is bonded to a pressed steel housing which contains a ball thrust bearing. Since the strut also acts as a telescopic damper, the inner rod carries a damper piston at its lower end. At the upper end of this rod a threaded portion of reduced diameter is locked to the rubber mounting

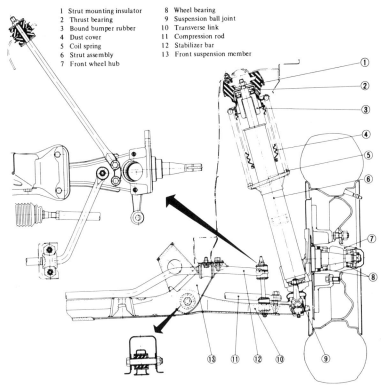

1 Strut mounting insulator
2 Thrust bearing
3 Bound bumper rubber
4 Dust cover
5 Coil spring
6 Strut assembly
7 Front wheel hub

8 Wheel bearing
9 Suspension ball joint
10 Transverse link
11 Compression rod
12 Stabilizer bar
13 Front suspension member

Fig. 5.16 MacPherson strut front suspension on Datsun 260Z.

and is located by the inner race of the ball bearing. This thrust bearing allows free angular movement of the tubular strut as the wheel is steered.

The Datsun is conventional in that the 'kingpin' offset is positive. Negative offset or *over centre-point steering*, as shown in Fig. 5.17, has been adopted by some designers of FWD cars since it was first used on the 1966 Oldsmobile Toronado to give improved stability under acceleration. Cars with dual. circuit braking often have brake wheel cylinders interconnected diagonally. Thus, if one braking circuit fails, one front brake and the diagonally opposite rear brake is still operative. With positive offset this type of failure creates a torque tending to turn the vehicle around the front wheel

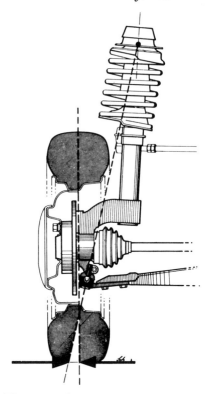

Fig. 5.17 Over centre-point steering.

with the effective brake. With hard braking this could result
in a spin. With negative offset the steering pull is in the
opposite direction, giving a better chance of straight-line
braking under partial brake failure. A very useful bonus is a
reduction in the danger of spinning when a front tyre blows
out at speed.

5.6 Independent rear suspension

Most innovators meet a wall of resistance when they present
their brainchild; even worse in their eyes, they meet
indifference. Inevitably, when the new ideas have proved

their worth, all the original objections evaporate, almost as if they had never existed. We saw it happen with IFS. With IRS, there has not yet been a complete victory. The advantages are indisputable, but the live beam axle when used at the rear is not as objectionable as when used at the front. The high unsprung mass is the only real disadvantage. Front wheel drive is becoming more popular. For the small family car it is becoming the norm. General Motors have designed a small 'world car' to be sold as the Astra by Vauxhall Motors, as the Kadett by Opel and, no doubt, under other badge guises in other countries. This follows the current trend with a transverse front engine and FWD.

On FWD cars therefore we see that the attractions of independent suspension at the rear is faced with the sheer simplicity of the dead beam axle, where the weight of a light axle tube and simple locating links is only slightly heavier than a good IRS system.

For a car of 1000 kg unladen weight, typical unsprung weights for five popular systems would be:

		kg/wheel
1	Live rear axle with coil springs and five-link location	= 50
2	de Dion axle	= 35
3	Double-wishbone IRS	= 25
4	Semi-trailing arm IRS	= 28
5	Dead rear axle	= 32

Typical values for the unsprung/sprung mass (m/M) ratio for the five systems would be:

	System	*F/R weight distribution*	*m/M*
1	(RWD only)	50/50	0.2
2	(RWD only)	50/50	0.14
3(a)	(FWD)	60/40	0.125
3(b)	(RWD)	50/50	0.10
4(a)	(FWD)	60/40	0.14
4(b)	(RWD)	50/50	0.112
5	(FWD only)	60/40	0.16

For a RWD car the superiority of IRS is indisputable, for a car with FWD the gain in road-holding appears to be marginal.

5.6.1   Wishbone IRS

The modern Jaguar rear suspension uses the drive shaft as an upper suspension link, as shown in the cross section of Fig. 5.18. Twin coil springs and telescopic dampers are used, one in front and the other behind the lower wishbone.

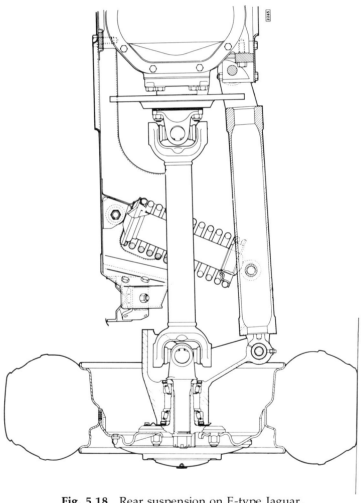

Fig. 5.18   Rear suspension on E-type Jaguar.

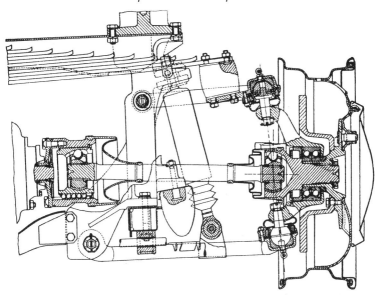

Fig. 5.19 Lancia Flavia constant-velocity joint incorporating rolling splines at inner end of drive shaft.

Longitudinal radius arms, not shown in the cross section, are attached to the hub carrier and resist acceleration and braking forces. By using a constant-length drive shaft, Jaguar cleverly avoid one of the design problems associated with double-wishbone IRS since sliding splines are not required. Plain splines can exert coefficients of friction as high as 0.25 when under power. This interferes with the free working of the suspension when under acceleration. Several early designs suffered from this problem, but the bearing manufacturers eventually provided an elegant solution. The Lancia Flavia of 1963 was an early example. Although this is a FWD car the principle applies equally to the case of RWD. It will be seen from Fig. 5.19 that the inner shaft universal joint uses rows of balls that are free to slide in semi-circular grooves arranged radially around the cylindrical housing. The outer shaft joint is a constant-velocity joint of the normal Rzeppa type. The Flavia used a transverse semi-elliptic spring which was typical of many early IFS designs. Variations in interleaf friction with this type of spring tended to sacrifice some of the gains given by the roller bearing splines. Coil springs and torsion bars have replaced semi-elliptic springs in modern designs.

5.6.2   Semi-trailing arms

A swing axle gives negative camber on the outer wheel under roll; a trailing arm gives positive camber. It is, therefore,

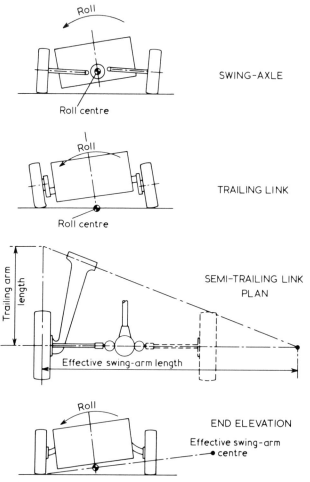

Fig. 5.20   Camber angle change under roll with swing axle, trailing link and semi-trailing link suspensions.

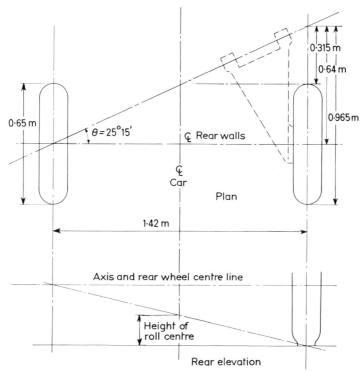

Fig. 5.21 Typical semi-trailing arm layout.

possible to combine them into a system called *semi-trailing arm* or *semi-trailing link* to give negligible camber change under roll. The concept was used on the Lancia Aurelia more than twenty-five years ago. The development of the concept is shown schematically in Fig. 5.20. The trail angle (ϕ, in Fig. 5.21) is chosen by the designer to suit the dynamic requirements of the particular vehicle. If for example he finds it necessary to increase cornering forces at the rear to give stable cornering, he could choose to increase the trail angle. Several of the current General Motors cars use a trail angle of 26°. In the example shown in Fig. 5.21 the trail angle of 24°15' gives an effective swing arm centre that coincides with the track. This layout would give a small amount of toe-in (20' of toe-in for a bump or rebound of 100 mm). Since toe-in

Fig. 5.22   Ford Granada semi-trailing arm suspension.

is produced on both outer and inner wheels, the roll-steer effect is negligible. When laying out a semi-trailing arm system, the designer has many parameters to consider. There is not only the obvious one of trail angle, there is the pivot axis, which is parallel to the ground in Fig. 5.21, but can be tilted if desired to change the static camber angle or can be positioned above or below wheel centre height. Finally, for every trail angle there is a choice of effective swing arm length: in Fig. 5.21 this is 1.42 m. If we increased this by 50% the roll centre height would be reduced by one-third and the toe-in angle for 100-mm bump or rebound would be reduced to about 13'. This would, of course, increase the length of the semi-trailing arm by 50%. This could be difficult to accommodate in the available space without intruding into the passenger area.

The Ford Granada rear suspension is shown in Fig. 5.22. Fabricated steel arms are used, pivoted on the cross-member that supports the final drive unit. The halfshafts carry a new design of constant-velocity sliding-type universal joint at both ends.

References

[1] Olley, M. (1934), 'Independent wheel suspension – its whys and wherefores', *SAE Journal*, March, p. 73.
[2] Goldman, E. D. and von Gierke, H. E. (1961), Effects of Shock and Vibration on Man, *Shock and Vibration Handbook*, vol. 3, McGraw-Hill, New York, chapter 44, p. 441.

6
Road-holding

Many years ago we defined road-holding as 'the ability of a car to follow the path dictated by an average driver in all reasonable circumstances'. An average driver on snow or ice will sometimes find that his car does not 'follow the path dictated'. Snow and ice, to an average driver, are not reasonable circumstances. To a Scandinavian rally driver, they are. With his skill and experience he will usually persuade the car to follow his wishes.

Depending upon the class of vehicle, we also demand the ability to negotiate a bend safely within the limits of tyre adhesion and the skill of the driver. Not only do we demand higher cornering speeds from racing cars, but we assume quicker reactions and greater skill from their drivers. The reverse is assumed in the case of normal road vehicles. If the driver should accidentally exceed the more modest cornering power of the tyres, the car should be designed to be 'forgiving' within the possible limits of engineering knowhow.

6.1 Straight-line stability

Oversteer and understeer were discussed briefly in Chapter 4. For straight-line stability a degree of understeer is essential. Let us consider a car travelling in a straight line when it is suddenly subjected to a side thrust from a gust of wind or from a dip in the road surface. Under the action of this side force, all four wheels will run at slip angles ϕ_f at the front and ϕ_r at the rear. If ϕ_f is less than ϕ_r, the car will steer towards the disturbing force. The car will now be following a curved path and since this will produce a centrifugal acceleration the initial slip angles ϕ_f and ϕ_r will be increased to new slip angles ϕ_f' and ϕ_r'. The increment to ϕ_f will be less than that given to ϕ_r, and this will again increase the centrifugal force to a higher value. The car is obviously in an unstable situation. This condition is called *oversteer* and, since frequent steering corrections are required to hold a car that oversteers in a straight line, it leads to rapid driver fatigue.

Stability can be produced by designing into the dynamic behaviour of the car larger slip angles at the front than at the rear under the action of transient side forces. With ϕ_f greater than ϕ_r, a centrifugal force is produced which is opposed to the side force. Within limits the car has intrinsic straight-line stability. Such a car is not necessarily stable when cornering.

6.2   Cornering dynamics

When a car is cornering, a much higher centrifugal force is produced than the minor side forces discussed in Section 6.1. To balance this force the tyres generate side thrusts, and to create these thrusts all four wheels must run at suitable slip angles. As will be seen in Fig. 6.1(a), the longitudinal axis of the car must be angled into the corner to achieve these angles. Although the steering geometry may be designed to give an Ackermann centre based on the rear axle centre line, the car could never turn about this centre since this would give zero slip angles at the rear. For simplicity, we have chosen to give identical steering angles to both inner and outer front wheels. This not only simplifies the vector diagrams in Fig. 6.1(b) and (c), but is a more logical steering geometry than that proposed by Ackermann.

The true turning centre will depend upon the average slip angles of all four tyres. In Fig. 6.1(a) an average slip angle of $7°30'$ has been chosen. In a typical understeering car this would probably involve an average of $8°$ at the front and $7°$ at the rear. As will be discussed later, one effect of body roll when cornering is to transfer load from the inner wheels to the outer. For any particular yaw angle, however, inner and outer slip angles are fixed by the steering and suspension geometry at the front and by the suspension geometry at the rear. It is, therefore, not unreasonable to take an average for the inner and outer slip angles in our simplified vector diagrams.

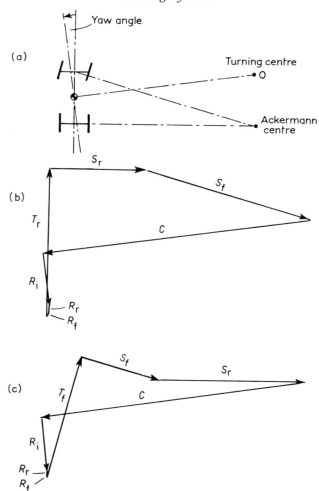

Fig. 6.1 Vector diagrams for cornering RWD and FWD medium-size saloon.

We have chosen a rather tight turning circle in the example of Fig. 6.1 to demonstrate the essential difference between the cornering forces associated with RWD and FWD. The data used in Fig. 6.1 represents a large European car or a 'compact' American car:

Total vehicle mass M = 1500 kg
Wheelbase   = 2.8 m
Track   = 1.6 m
Turn radius r = 10 m
Velocity v = 7 m/s.

From this, we obtain:

$$\text{Centrifugal force } C = \frac{Mv^2}{r}$$

$$= \frac{1500 \times 7^2}{10}$$

$$= 7350 \text{ N.}$$

This is a centrifugal force of about 0.5 gravity.

Fig. 6.1(a) is drawn to scale with a yaw angle of $7°30'$. This yaw angle is necessary to promote the required slip angles. The turning radius is perpendicular to this instantaneous yaw axis and the turning centre is at 0.

6.3   Rear wheel drive

In the Circle of Forces or closed vector diagram for the RWD car in Fig. 6.1(b) the centrifugal force C of 7350 N is balanced by the following forces:

S_r, the cornering force from the rear tyres;
T_r, the tractive force from the rear tyres;
S_f, the cornering force from the front tyres;
R_i, the inertia force required to produce acceleration of the mass through the bend;
R_f, the road resistance of the front wheels;
R_r, the road resistance of the rear wheels.

If we refer to our earlier concept of a tyre-footprint Circle of Forces (see Fig. 1.17 in Chapter 1), we see that T_r and S_r can be found by the following relationship:

$$\text{Total tyre thrust} = \sqrt{(S_r^2 + T_r^2)}.$$

If we assume for simplicity that the front and rear tyres exert identical *total thrusts*

$$S_f = \sqrt{(S_r^2 + T_r^2)}$$

R_i, the inertia force, is that force required to accelerate the mass through the bend:

$$R_i = Ma$$

where a is the acceleration. In the particular example a value for a of 1.0 m/s^2 has been taken.

Therefore $\qquad\qquad R_i = 1500$ N.

Although the speed is low, the bend radius is enough to increase the road resistance to an appreciable value. A value of 75 N per wheel seems reasonable. No great accuracy is required in estimating this value. If the values R_r and R_f of 150 N for each pair of wheels were neglected, the error would not be great.

Only C, R_i, R_f and R_r are known in direction and magnitude. T_r, S_r and S_f are known in direction only, but the relationship $S_f = \sqrt{(T_r^2 + S_r^2)}$ helps us to construct a closed vector diagram by trial and error, as shown in Fig. 6.1(b).

6.4 Front wheel drive

The same technique has been used to construct a closed vector diagram for the case of a FWD car. It is immediately apparent that the direction of the tractive effort in this case has a centripetal component that opposes a percentage of the centrifugal force.

Scaling off values from Fig. 6.1(b) and (c), we get the following comparison:

	RWD	FWD
S_f	4700 N	2200 N
S_r	2650 N	4050 N
T_r	3950 N	
T_f		3400 N

A direct comparison of the cornering 'efficiency' of the two systems can be obtained by comparing the value of S_f in Fig. 6.1(b) and S_r in Fig. 6.1(c). This shows a reduction in cornering forces of about 15% in the second case when resisting the same centrifugal forces. For the same limiting slip angles the FWD car should be able to negotiate such a tight corner at about 7% higher speed. This gain falls to about 3% for a 20-m radius corner and becomes negligible for large radius curves.

6.5   The cornering racing car

Modern racing cars carry such a high percentage of the weight and the footprint area at the rear that they can be said to be steered in a bend by the throttle foot. A typical weight distribution would be 33/67 and typical tyre widths 20 in at the rear and 10 in at the front. Rear end loads are usually increased in all but the most acute bends by the downthrust of a rear wing, a large inverted aerofoil. A smaller front wing is provided to increase the effective load on the front wheels. During 1978–9 the overall downthrust was increased beyond the actual weight of the car. This was achieved by the application of a Venturi-effect underneath the car and the development of sliding skirts on the body sides to 'seal-in' the depression below the car. The penalty of these aerodynamic tricks has been a loss in maximum speed, but the cornering speed has been increased dramatically. Measurements reported by *Motor* (3 March 1979) made on a Lotus 79 gave cornering accelerations of 2.05g. Fig. 6.2 shows how the downforces are distributed on the Lotus 79.

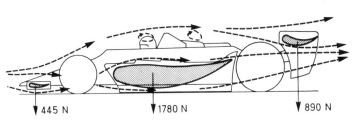

Fig. 6.2   Aerodynamic downforces on F1 racing car at 150 m.p.h.

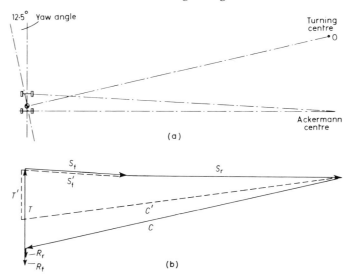

Fig. 6.3 (a) and (b): Vector diagram for cornering racing car.

6.5.1 Cornering dynamics

Fig 6.3(a) shows a typical Formula 1 car at a yaw angle of 12°. The following data have been used to construct the vector diagram of Fig. 6.3(b):

Total vehicle mass M	= 725 kg
Wheelbase	= 2.7 m
Track	= 1.6 m
Weight distribution F/R	= 33/67
Turn radius r	= 50 m
Velocity v	= 30 m/s

Therefore,

$$\text{Centrifugal force } C = \frac{Mv^2}{r}$$

$$= \frac{725 \times 30^2}{50}$$

$$= 13050 \text{ N.}$$

This is a centrifugal force of about 1.8g.

In Fig. 6.3(b) the broken lines C', T' and S'_f show how the driver can lift his throttle foot, reduce the torque from the rear wheels and change the turning centre. This, of course, is a steering action. The changes in S_f and S_r are negligible for a large change in direction, giving the car good stability in a corner.

6.6 Yaw inertia

In Chapter 3 the *polar moment of inertia* about a horizontal axis through the centroid was considered. This has an influence on the natural period in pitch. The disposition of the major

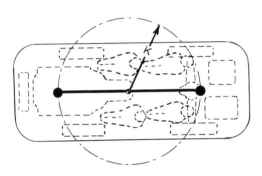

(a) HIGH POLAR MOMENT OF INERTIA

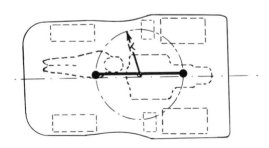

(b) LOW POLAR MOMENT OF INERTIA

Fig. 6.4 How the distribution of the major masses influences the radius of gyration.

masses in the car also influences the resistance of the vehicle to change in direction. This polar moment of inertia is that of the whole vehicle, including the unsprung components and is a measure of the car's resistance to rotation about a vertical axis through the centroid. Since this rotation involves a change in yaw angle, it is sometimes called the *yaw moment of inertia*.

Racing drivers discovered many years ago that long-wheelbase versions of the same basic racing car were much more stable at speed than the short-wheelbase model. This was before the days of aerodynamics. When negotiating a twisting part of a round-the-houses circuit, they preferred the handling of the short-wheelbase version. Fig. 6.4 shows how a front-engine rear drive car, carrying a full complement of passengers, will have a much higher yaw moment of inertia than a mid-engine sports car with only the driver aboard.

The Moment of Inertia in Yaw is given by:

$$I_y = M_t K_y^2 \qquad (6.1)$$

where M_t is the total mass, and K_y is the radius gyration of the complete car. The couple, to produce any change in direction, must come from the tyre footprints acting through effective arms of length l_1 at the front and l_2 at the rear.

If we consider the vehicle as changing from the straight-ahead direction to an angular velocity of ω rad/s in t seconds, the angular acceleration $\alpha = \omega/t$ rad/s^2.

The tyre forces to produce this angular acceleration are:

$$\Sigma s = \frac{I_y \alpha}{l_1 + l_2} . \qquad (6.2)$$

In the example of Fig. 6.1 the angular velocity

$$\omega = \frac{v}{r} = \frac{7}{10} = 0.7 \text{ rad/s}.$$

If this rate of angular velocity is achieved in 1 s, the angular acceleration would be 0.7 rad/s^2.

A typical value of I_y for a medium-size saloon of 1500 kg would be 2500 kg/m^2. The summation of the tyre forces (inwards at the front and outwards at the rear) would be:

$$\Sigma s = \frac{2500 \times 0.7}{2.8} = 620 \text{ N.}$$

Dividing this equally between front and rear in the example chosen for the vector diagram in Fig. 6.1(b), the front tyre forces would be increased from 4700 N to 5010 N and the rear tyre forces reduced from 2650 N to 2340 N. These are only transient forces induced during the 1 s duration while the car is steered from the straight-ahead direction into the 10-m radius curve of Fig. 6.1(a). Since this only represents a centrifugal force of about $0.5g$, the transient angular acceleration would be easily accommodated on a dry road by an increase in the front slip angles and a reduction at the rear. On a wet road the sudden application of $0.5g$ could provoke a front wheel skid.

The above analysis of the influence of yaw inertia neglects the effects of roll-steer and load transfer from inner to outer tyres under the effects of roll. Some modern sports cars introduce another variable into this complex situation. The Lotus Esprit, for example, uses 205/60 HR14 tyres at the front at a pressure of 18 p.s.i. (124 kN/m^2) and 205/70 HR14 at the rear at a pressure of 28 p.s.i. (193 kN/m^2). The 60 profile tyres at the front have a larger footprint area than the 70 profile tyres at the rear and will, therefore, transmit greater side forces for a given slip angle.

6.6.1 Spin or slide?

If the driver of a very large four-wheeled vehicle such as a bus tries to change direction too quickly on a slippery surface, the breakaway usually occurs at the front end first. Once the vehicle is sliding, however, the angular momentum that provoked the slide usually results in a spin; the larger the ratio $K_y^2/(l_1 \times l_2)$, the more prolonged the spin. A small car like the Volkswagen Beetle with a tail-heavy weight distribution usually begins to slide at the rear under similar circumstances and a spin is inevitable with a driver of less than average ability or even a good driver who has been behind the wheel for too long.

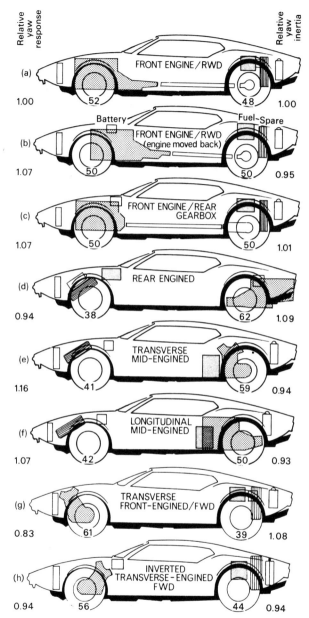

(a)   FRONT ENGINE / RWD

1.00   52   48   1.00

Battery   Fuel   Spare

(b)   FRONT ENGINE / RWD
(engine moved back)

1.07   50   50   0.95

(c)   FRONT ENGINE / REAR
GEARBOX

1.07   50   50   1.01

(d)   REAR ENGINED

0.94   38   62   1.09

(e)   TRANSVERSE
MID-ENGINED

1.16   41   59   0.94

(f)   LONGITUDINAL
MID-ENGINED

1.07   42   50   0.93

(g)   TRANSVERSE
FRONT-ENGINED/FWD

0.83   61   39   1.08

(h)   INVERTED
TRANSVERSE-ENGINED
FWD

0.94   56   44   0.94

Fig. 6.5 Relative yaw responses (on left) and relative yaw inertias (on right) for eight possible mass distributions on a modern sports car.

6.6.2   Yaw responses

Measurements have been made by various research institutions of the influence of yaw inertia on the response of a car to a given steering input. *Yaw response* is usually taken as the lag between the input from the steering wheel and the output in terms of change in angular momentum.

Fig. 6.5 is based on a series of calculations made by the Jaguar Car Co. and presented in a paper to the SAE Detroit Congress (Feb.–Mar., 1977) by Bob Knight in conjunction with Jim Randle. The particular study was made as a justification for the layout chosen for the XJ-S. Not that it needs any! In Bob Knight's comparison yaw response is the inverse of the above definition. It is the rate of change of angular momentum for a given change in steering angle. In Fig. 6.5, (a) is taken as unity. A car with a higher value is, therefore, one which in common language 'responds more quickly to the helm'. None of the examples can be said to represent the XJ-S which has a weight distribution of 57/43 and has a different profile to those of (a) and (b).

As one would expect, the poorest yaw response is given by (g). Not only is the value of I_y 8% higher than that of (a), but the weight distribution of 61/39 will obviously make the front slip angles increase to a greater extent for a given steering angle change than would occur with a tail-heavy car. In contrast we see that the tail-heavy (e) with a relatively low yaw inertia of 0.94 has the highest relative yaw response. With a weight distribution of 41/59, however, great care will be necessary in the layout of the suspension system to avoid the dangers of terminal oversteer when cornering near the limit.

The above analysis suggests that the range of yaw responses for a comprehensive list of modern sports car designs is not large. If we extended the survey to cover everything from a small sports car such as an MG Midget to a full-size American sedan, we would find the range of yaw responses to be much greater; certainly greater than 2 to 1.

6.6.3   Spin accelerations

A recent attempt by the staff of *Motor*, with the help of the Cranfield School of Automotive Studies, to measure the maximum spin accelerations that could be reached on a wetted skid pad produced some conflicting results. A Lotus Elite, with a relatively high yaw moment of inertia of 2006 kg/m^2 and a value for $K_y^2/(l_1 \times l_2)$ of 0.96, gave an almost identical spin acceleration to the mid-engined Esprit for which the corresponding values are 1508 kg/m^2 and 0.83. The Elite has 205/60 tyres all round, while the Esprit has 205/70 at the front and the wider 205/60 at the rear. Such differences as profile ratios are obviously more important than differences in yaw inertias.

6.7 Roll-steer

When a wheel is deflected from its normal axis of rotation by body roll, it can influence the car's direction. When both wheels at the front or both wheels at the rear are deflected by roll, the roll-steer effect can be quite pronounced. It all depends upon the suspension geometry.

An example of roll-steer can be illustrated by the simple example of a dead rear axle (see Fig. 6.6) as used on many FWD cars in which the trailing links slope upwards to the rear. As the body rolls, the more heavily loaded outer wheel moves upwards and *forwards* while the inner wheel moves downwards and *backwards*. This produces understeer. In practice the link pivot centre would be only slightly below wheel centre height, probably no more than 50 mm. The same technique can be used with a semi-trailing link suspension or with any independent suspension design using longitudinal links to control the fore and aft location of the wheels. It can be applied to both front or rear suspensions.

Excessive roll oversteer can make a car 'twitchy' on an uneven road surface. A memorable example was the original Mercedes-Benz 300SL with double-wishbone suspension at the front and swing axle suspension at the rear. The low roll centre at the front, the very high roll centre at the rear, plus the variable cornering power produced by single wheel bump and rebound at the rear as the rear wheels alternate between positive and negative camber, all contribute to a lack of stability of speed. (I myself remember this tendency to weave from side to side at high speed!) Daimler-Benz removed this inherent instability when they introduced the low-pivot swing axle design.

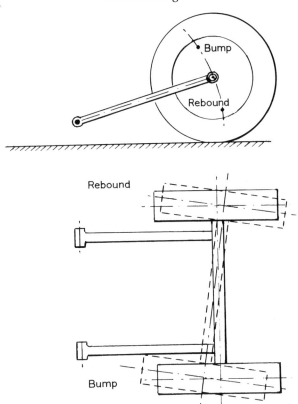

Fig. 6.6 Deflection steer: roll understeer on a dead rear axle.

6.7.1 Deflection steer

On production cars it is common practice to use rubber bushes at the suspension pivots. With metal-to-metal contact far too much harsh road vibration would be transmitted to the occupants. Unfortunately, the deflections that can occur under changing loads applied to these flexible pivots can produce undesirable changes in the suspension and steering geometry. Even variations introduced by production tolerances, can give problems. The original Ford Escort was designed to have zero roll-steer on the front wheels. In practice some cars leaving the assembly line toed-in under

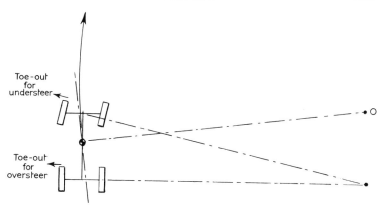

Fig. 6.7 Effects of toe-out deflection changes at front and rear.

bump (oversteer), while others toed-out (understeer). When the front suspension was modified in 1968, the geometry was changed to give roll understeer at the front on all cars. The designed roll understeer was 13 minutes per degree of roll. This would be greater on some cars and less on others, but the manufacturing tolerances were such that no cars would reach the customer with a tendency to roll oversteer.

Deflection steer can occur at either end of the car. As shown in Fig. 6.7, toe-out of the outer wheel under roll produces understeer at the front but oversteer at the rear. The Porsche engineers who designed the type 928 tackled the problem of deflection steer in a novel way. This patented 'Weissach-Axle' is described in Chapter 12, in which we present a detailed analysis of the type 928 suspension.

7
Dampers

7.1 Friction dampers

Originally called *shock absorbers*, dampers were first introduced in the 1920s as a palliative to reduce excessive movements of the front axle. The popular André-Hartford friction damper, shown in Fig. 7.1, uses a sandwich arrangement of alternating flat disks of steel and friction material, the whole assembly tightened by an adjustable dished spring. Two levers in V-formation, pivotally attached to the axle and the chassis frame, produce relative angular movement between the steel disks and the friction disks, thus providing a damping action.

A serious disadvantage of the friction damper was a tendency to exert high initial or starting friction, i.e. more friction was required to create initial movement than was exerted when moving. The model shown in Fig. 7.1 is the Telecontrol adjustable shock absorber. In this design the pressure on the friction disks could be varied by means of a Bowden cable controlled by a lever on the dashboard.

7.2 Hydraulic dampers

A simple hydraulic damper can be made from a piston working inside a cylinder with oil on both sides of the piston.

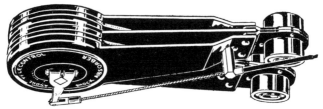

Fig. 7.1 André-Hartford Telecontrol shock absorber.

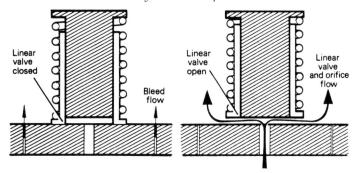

Fig. 7.2 The principle of the 'ride control' or 'linear valve'.

A series of holes in this piston and a spring-loaded valve allow oil to flow from one side of the piston to the other, the size and number of holes and the design details of the valve determine the forces exerted by the damper on the suspension system. The piston is usually attached to the

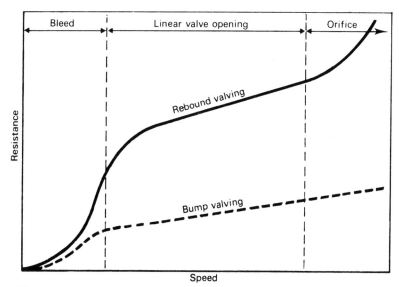

Fig. 7.3 Typical flow characteristic curves given by bleed flow and linear valve flow.

body, the cylinder to the unsprung mass. It is easy to see that the use of a viscous fluid in the damper will give a damping force that is 'velocity-dependent', i.e. $F \propto V^n$. The value of the index n will depend upon the design of the spring-loaded valve. At very low vertical wheel velocities the valve is designed to remain closed, all oil flow being through the bleed orifices. These are small enough to give streamline flow. The modern 'ride-control' or 'linear' valve is spring-loaded, as shown in Fig. 7.2. By changes in spring strength and the diameter and port dimensions of the valve orifice, a wide range of resistance/flow curves are possible. As the linear valve opens under high fluid flows, a point is eventually reached when the valve disk is lifted so high that the flow is controlled by the diameter of the valve orifice. This gives a gradually increasing rate of rise in resistance, as seen in the rebound valving curve of Fig. 7.3.

Most damper designers provide different valves for bump and rebound flow, the resistance usually being much less in bump than in rebound. Typical characteristic resistance curves are given in Fig. 7.3.

7.3 The double-tube damper

Older designs of telescopic damper used two concentric tubes. The inner tube was the working cylinder, the outer tube acting as a reservoir to hold the excess oil moved on the downstroke. A good example is the Koni, shown in cross section in Fig. 7.4. Since liquids are almost incompressible, the necessity for the outer chamber is apparent when it is seen that the effective area on the upstroke is less than that on the downstroke by the cross-sectional area of the piston rod.

7.4 The single-tube damper

The Girling Monitube damper uses a principle first used by de Carbon nearly fifty years ago. With modern shaft-sealing

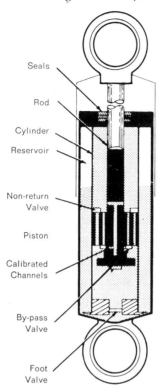

Seals
Rod
Cylinder
Reservoir
Non-return
Valve
Piston
Calibrated
Channels
By-pass
Valve
Foot
Valve

Fig. 7.4 Koni double-tube damper.

techniques, Girling has produced a successful development of
this old idea. The recuperation cylinder in this damper is an
extension of the working cylinder containing gas under high
pressure. A lightweight glass-reinforced, nylon-free piston is
used to isolate the two chambers. Movement of this light
piston accommodates the displacement differences of the
working piston. Valves for both bump and rebound are
incorporated in the working cylinder, but are omitted from
the simplified diagram in the drawing on the left of Fig. 7.5.

Another ingenious single-tube damper has been developed
by Woodhead-Munroe, shown in principle on the right of

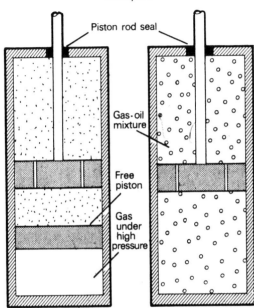

Piston rod seal

Gas·oil mixture

Free piston

Gas under high pressure

Fig. 7.5 Girling Monitube principle (left) and Woodhead-Munroe Monotube principle (right).

Fig. 7.5. Aeration in earlier designs of hydraulic damper sometimes occurred when they had been subjected to heavy-duty work, such as a long drive over rough roads. The effect of aeration was similar to a large reduction in oil viscosity with a corresponding drop in the effectiveness of the dampers. The Woodhead-Munroe engineers decided that the most elegant solution to the problem was to keep the oil aerated at all times. By charging the damper cylinder with a finely emulsified mixture of oil and gas, they also solved the problem of the differing swept volumes on bump and rebound, since the variations in displacement could be absorbed by compression of the tiny gas bubbles. The valving had to be designed to suit the liquid/gas mixture, since this had a relatively low viscosity, but this had now become an effective viscosity that varied very little from the start to the end of a journey. A cross section of a Woodhead-Munroe monotube damper is given in Fig. 7.6.

BUMP   REBOUND

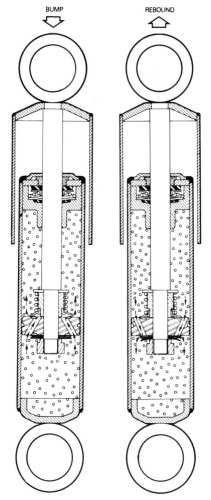

Fig. 7.6 Woodhead-Munroe Monotube damper in cross section.

7.5 Damping theory

Although our aim is to damp out as quickly as possible each vertical impulse given to the wheels, a measure of compromise is necessary. Excessive damping gives a reduction in the initial spring deflection. This gives a harder ride.

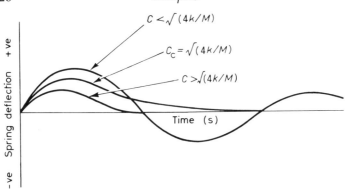

Fig. 7.7 Damping with various damping coefficients.

7.5.1 Critical damping

The concept of a *critical damping coefficient* is a purely mathematical one and must not be confused with the more practical concept of *optimum damping*.

Critical damping is given when the damping force is the *minimum* force required to prevent the initial displacement of the sprung mass·from the equilibrium position, resulting in an oscillation that crosses the equilibrium position. Critical damping is indicated in Fig. 7.7 by the curve labelled C_c.

The critical damping coefficient is given by:

$$C_c = \sqrt{\frac{4k}{M}} \qquad (7.1)$$

where k = the spring rate (N/m) and M = the sprung mass (kg). The damping coefficient has the units of frequency, i.e. $1/s$.

The mathematics involved in establishing the above equation are very involved. Students are referred to *Mechanics of Road Vehicles* by W. Steeds [1] for confirmation. Critical damping, in any case, is much too severe for a practical suspension system, but it does give us a yardstick by which we can express the degree of damping given to a practical system. On this scale most modern automobiles are given damping coefficients C where $C/C_c = 0.25$ to 0.5.

A typical modern damper would damp out the initial bump

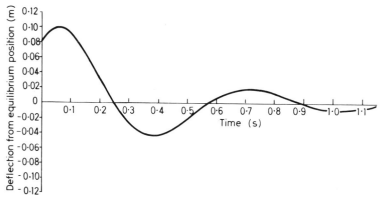

Fig. 7.8 Attenuation curve for typical damper.

deflection of x to a value of no more than $0.2x$ by the second bump in a time interval of $1/F_n$ seconds, where F_n is the natural frequency of the suspension. This is illustrated in Fig. 7.8.

7.5.2 Resonance

We have already referred to the phenomenon of resonance in Chapter 2. As shown in Fig. 2.6, with a complete lack of damping the effect of a regular wave formation in the road surface could be quite shattering when the forcing frequency f_r coincides with the natural spring frequency f_n. Curve 1 demonstrates this. Curve 2 with a maximum acceleration at resonance of about $0.7g$, has a value of $C/C_c \simeq 0.15$. Curve 3 is for a value of $C/C_c \simeq 0.25$ and Curve 4 is for $C/C_c \simeq 0.65$. Curve 3 has been labelled 'well-damped', but these curves are based on a damping function directly proportional to vertical wheel velocity. Damper designers have accumulated a wealth of practical experience in the design of ride-control valves that gives a finished product giving good damping over a very wide range of f_r/f_n values.

7.5.3 A mathematical model

Before we can hope to use computers to design suspension systems, we must first establish reliable mathematical models.

This was an irresistible challenge to academic minds, even before computers were available. As early as 1950 de Carbon presented a paper to the Congrès Technique Internationale de l'Automobile in Paris on the theory and practice of vehicular suspension damping. After thirty years of mathematical analysis, reasonably accurate models can be constructed to help the design engineer through his computer facilities to optimize the suspension damping on a new model. (One textbook and four useful technical papers, [2], [3], [4], and [5], to assist the serious student are given in the References.)

References

[1] Steeds, W. (1960) *Mechanics of Road Vehicles*, Iliffe & Sons, London.
[2] Bender, Karnopp and Paul (1967), 'On the optimization of vehicle suspensions using random process theory', *Proc. 1967 Transportation Engng Conf.*, ASME paper 67-trans-12.
[3] Thompson, A. G. (1969–70), 'Optimum damping in a randomly excited non-linear suspension', *Proc. Inst. Mech. Engrs*, (AD) 1969–70, vol. 184, pt. 2A, no. 8, pp. 169–84.
[4] Thompson, A. G. (1972), 'A simple formula for optimum suspension damping', *J. Automotive Engng, Inst. Mech. Engrs*, April.
[5] Hever, P. J. (1969), 'Interaction of suspension and vibration damping in road vehicles', *ATZ*, **71**, (12) (December), pp. 446–51 (in German).

8

Pneumatic Suspensions

8.1 The air spring

A mechanism for compressing and expanding an enclosed volume of air by means of a piston or diaphragm was first patented as a suspension spring in 1906 and a Cowey car fitted with pneumatic suspension was exhibited at Olympia Motor Show in 1909.

To demonstrate the concept we can compare the behaviour of a conventional steel spring, such as a coil spring or a torsion bar and that of a pneumatic strut, shown schematically in Fig. 8.1. Friction in the strut will be

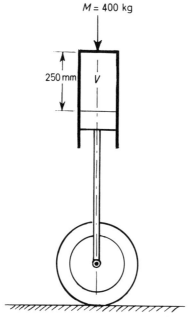

Fig. 8.1 Elementary pneumatic strut.

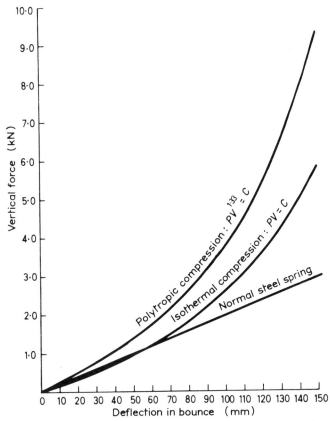

Fig. 8.2 Comparative compression forces with normal coil spring and pneumatic spring.

neglected. Let us consider the mass supported by the spring as 400 kg. With a rate of 20 kN/m a coil spring of constant gauge, coil diameter and helix angle will exert a restoring force of 0.5 kN for a deflection of 25 mm, 1.0 kN for a deflection of 50 mm and pro rata up to a value of 3.0 kN for a deflection of 150 mm, as shown in Fig. 8.2.

Very slow compression of the air in the strut would give a rate curve following the law of isothermal compression, i.e. PV = a constant. At speed however on a typical road surface, the frequencies imposed on the suspension system, omitting those absorbed by the tyre, vary from zero to about 40 Hz.

Compression and expansion in general will, therefore, be very rapid and will be much closer to adiabatic. Since some cooling will inevitably occur, polytropic compression and expansion is assumed with an index of $n = 1.33$. This curve in Fig. 8.2 is, therefore representative of the compression curve for a simple pneumatic strut where $PV^{1.33} = C$.

From Fig. 8.2 we see that the spring rates for small deflections are fairly close, but for large deflections the pneumatic spring has a much greater rate. The extent of this increase in rate can be varied by increasing or decreasing the strut length in relation to the maximum deflection. Alternatively, an auxiliary chamber can be provided to increase the compressed volume at maximum bump.

8.2 Temperature variations

A simple pneumatic strut would give a variable static ride height. If the ride height was at the designed height of Fig. 8.1 at 20°C, it would fall by 34 mm if the car stood overnight at a temperature of −20°C. More devastating would be the effect of a rise in temperature from compression heating as the vehicle travelled on a rough road. A calculation made in *Automobile Engineer* in March 1955 suggested a rise of about 65°C in the air temperature without the provision of cooling fins or a forced cooling system. With an ambient temperature of 35°C this would increase the ride height by 68 mm, an unacceptable increase.

8.3 Self-levelling systems

Variations in ride height is something we have lived with for so long that we accept it as normal. Consider a typical family saloon with a load on each rear spring of 300 kg with an empty rear seat and nothing in the boot (trunk). With a typical rear spring rate of 20 kN/m and an available wheel travel in bump of 120 mm (4.7 in) the rear ground clearance could be as shown in Fig. 8.3, i.e. 175 mm (6.9 in). An increase in load of three passengers at the rear and 50 kg of luggage would increase the load on each rear spring by about

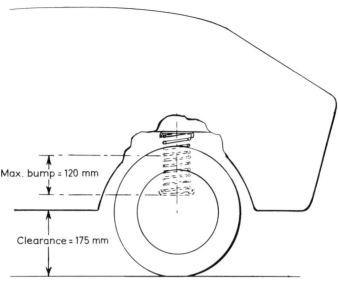

Fig. 8.3

140 kg. The additional downforce on each rear spring would be 140 × 9.807 = 1373 N, giving an additional static spring compression of 69 mm (2.7 in). The static ground clearance at the rear would be reduced to 106 mm (4.2 in) and the available spring travel in bump reduced from 120 mm (4.7 in) to 51 mm (2.0 in). On a rough road the rear springs would spend much of the time hitting the bump stops.

A self-levelling system is an obvious answer to carrying a wide range of loads. What is of no less importance is the freedom it gives to the designer to soften the suspension. With a bump travel of 120 mm available under all load conditions the spring rate can be softened from 20 kN/m to as low as 12 kN/m. This would reduce the rear spring frequencies from the typical conventional values of 1.3 Hz (2-up) and 1.07 Hz (5-up) to 1.0 Hz (2-up) and 0.83 Hz (5-up).

8.3.1   Practical air suspensions

The Firestone Tire Co. of America experimented in the 1930s with bellows springs, using two convolutions of the type

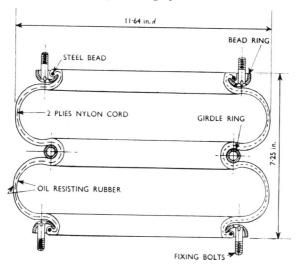

Fig. 8.4   Section through two-convolution air spring.

shown in Fig. 8.4. It was a rubber and fabric moulding with a steel bead ring at each end and a steel girdle ring to resist expansion in the middle. When the upper and lower bead rings were bolted to an upper steel plate attached to the body, and a lower steel plate mounted on the lower suspension arm, an enclosed chamber was formed to act as an air spring.

From these early experiments the company developed an air suspension system which was fitted to long-distance Greyhound coaches in the 1950s. In this installation the bellows unit was designed to operate at a pressure of 4.5–5.2 bar (65–75 lbf/in^2). Since commercial-vehicle air brakes in the USA are designed to operate at slightly higher pressures, this made the system suitable for use with the standard air brake compressor. Natural frequencies in the range 1.3–1.5 Hz could be attained using this system. These were very low frequencies for a large bus. Since the bellows unit was frictionless, apart from the negligible amount of hysteresis in the moulded bellows, it was necessary to provide good double-acting dampers and bump stops to cope with the worst roads.

In the Greyhound installation each suspension unit

consisted of a platform mounted above the axle with a bellows unit fore and aft. A surge tank was installed above each pair of bellows. The bellows had no inherent stability, but were located longitudinally by radius rods and laterally by a Panhard rod. The total volume of the air spring was that of the two bellows units plus the capacity of the surge tank, since an unrestricted flow was permitted between the three units. The action of the surge tank was to control the spring rate and the natural frequency, as will be described later.

From each surge tank a pipe passed to a levelling valve, there being one levelling valve per wheel. The levelling valves were pressurized from the main storage tank which was in series with the brake system storage tank. A check valve, fitted between the two tanks, ensured an air supply to the brakes if the suspension system failed. The levelling valves maintained a constant ride height on all wheels irrespective of the static load. The valve body was secured to

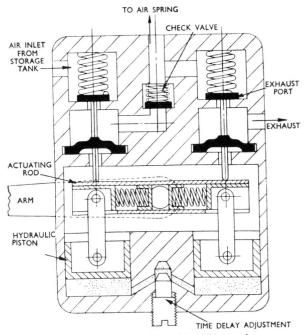

Fig. 8.5 Cross section of levelling valve.

the chassis frame, the valve actuating arm to the axle. An increase in load moved the arm inwards (see Fig. 8.5) which in turn tilted the actuating rod to open the inlet valve and admit compressed air to the air spring. Conversely, with a reduction in load the actuating rod would tilt in the opposite direction to exhaust air from the spring until the actuating arm became level again. Valve operation was not instantaneous, however, since a time-delay mechanism was built into the system. The end of the actuating arm acted on the rod through balanced compression springs contained in a central passage in the rod. These in turn were attached by vertical links to two hydraulic dampers. The speed of movement of the actuating rod was controlled entirely by the time-delay bleed between the two dampers. With a sudden movement of the actuating arm in either direction tilting of the rod only occurred after several

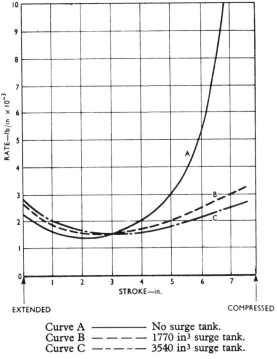

Curve A ——————— No surge tank.
Curve B — — — — 1770 in³ surge tank.
Curve C — — - — — 3540 in³ surge tank.

Fig. 8.6   Effect of surge tank volume on spring rate.

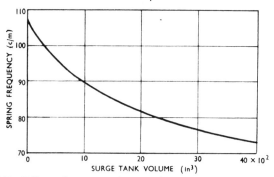

Fig. 8.7 Effect of surge tank on volume spring frequency.

seconds, the actual time being controlled by the time taken
for the appropriate compression spring to overcome the
resistance to flow through the orifice.

The time delay was essential to desensitize the levelling
valves. With no time delay the valves would open and close
under the action of every bump and rebound movement. Not
only would this put a high demand on the air supply and
require the provision of a very large compressor, but the
concept of four individual air springs would be upset and
fore-and-aft and side-to-side interaction could occur.

The actual delay time was established on the Greyhound
buses by trial and error. It was later found entirely
satisfactory to fit two levelling valves to the rear axle and a
single valve at the front to control both front springs. The
valve was designed to prevent any transfer of air from side to
side.

Without a surge tank the spring rate increase with
increasing wheel movement would have been far too drastic,
as is shown in Fig. 8.2. On such a large vehicle a large surge
tank could easily be accommodated, but a large-bore
connecting pipe was needed to ensure that the two bellows
and their surge tank would behave as if they were a single
large volume. Fig. 8.6 shows how spring rate changes could
be reduced by the addition of a surge tank. Fig. 8.7 shows
how spring frequency is reduced as the surge tank volume is
increased.

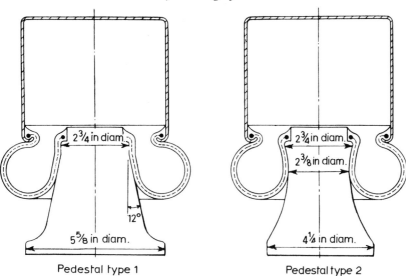

Pedestal type 1   Pedestal type 2

Fig. 8.8 Comparison of pedestal shapes.

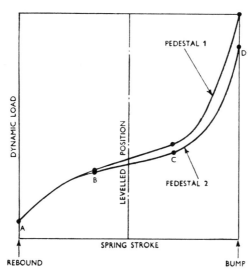

Fig. 8.9 Dynamic load curves using pedestals of different profiles.

8.3.2   Single-convolution air springs

With the automobile market in view the Firestone Co. developed a rubber and canvas bellows with only one convolution. To vary the spring rate characteristics, curved or tapered pedestals were developed to give an effective piston area that changed throughout the stroke. Fig. 8.8 shows two such pedestals, and Fig. 8.9 shows how a change in pedestal profile can influence the load/deflection curve. With pedestal 2 soft suspension will be given for small deflection in both bump and rebound (curve BC). For greater deflections in bump the spring operates over the rising section (CD). In rebound the spring operates over a more modest rising rate curve (BA).

8.4   The General Motors air suspension

The Eldorado Brougham was introduced in 1957 with air suspension by the Cadillac Division of General Motors. Despite the excellent ride and the advantages of self-levelling, the system was replaced by conventional steel coil springs in 1960. Problems with air leaks in some components and frozen valves in subzero conditions would undoubtedly have been solved by the Cadillac engineers, given time, but the

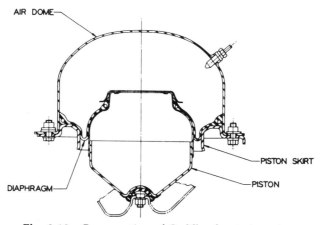

Fig. 8.10   Cross section of Cadillac front air spring.

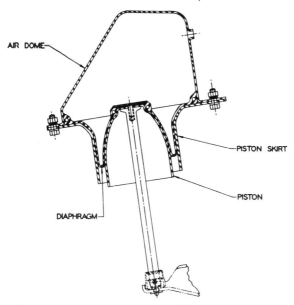

AIR DOME

PISTON SKIRT

PISTON

DIAPHRAGM

Fig. 8.11 Cross section of Cadillac rear air spring.

proven reliability of coil springs was an inexpensive and quick solution to an embarrassing situation. It was admitted by *Motor* that the earlier Cadillacs had set a very high standard. Even so they stated that the air-sprung Eldorado 'gave an amazing improvement' over badly surfaced roads.

On this GM design each front surge tank was mounted neatly inside the frame side-member, as shown in Fig. 8.10. The bellows unit became a compact design of rolling diaphragm (see Fig. 8.10). This diaphragm had a 3-in diameter hole in the centre, thus making the volume below the diaphragm part of the effective volume.

The rear axle was located by a four-link system and suspended on an air spring at each side. Domed pistons or pedestals, shown in cross section in Fig. 8.11, gave a rising rate suspension in bump and an almost constant rate in rebound. With more space available a large surge tank could be provided at the rear.

The system was designed to operate with an air pressure of 75 lbf/in^2 (5.17 bar). Spring frequencies were approximately

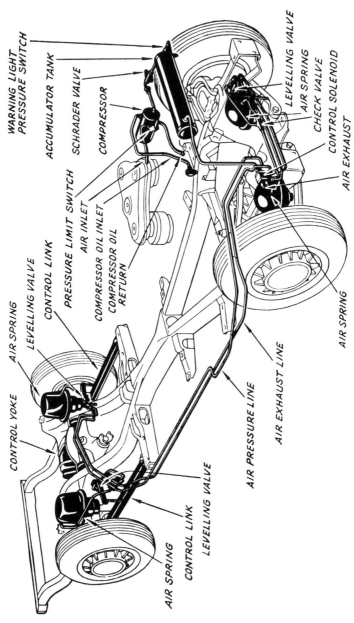

WARNING LIGHT
PRESSURE SWITCH

ACCUMULATOR TANK

SCHRADER VALVE

COMPRESSOR

PRESSURE LIMIT SWITCH

AIR INLET

COMPRESSOR OIL INLET

COMPRESSOR OIL
RETURN

LEVELLING VALVE

AIR SPRING

CHECK VALVE

CONTROL SOLENOID

AIR EXHAUST

AIR SPRING

AIR EXHAUST LINE

AIR PRESSURE LINE

LEVELLING VALVE

CONTROL LINK

AIR SPRING

CONTROL LINK

LEVELLING VALVE

AIR SPRING

CONTROL YOKE

Fig. 8.12 Layout of pneumatic suspension on Cadillac Eldorado.

0.9 Hz, or about 10% lower than the later models with coil springs. As on the earlier bus design an air compressor, plus an automatic levelling system, was provided.

8.4.1 Compressor and accumulator

An electrically driven piston-type air compressor with a delivery of 600 in³/min (0.59 m³/h) was mounted above the generator in the engine compartment. The compressor operated upon demand from a pressure switch on the 500 in³ (8200 ml) accumulator. This vessel also served as a trap for oil or condensed water and was provided with a manually operated drain valve. A warning light on the instrument panel informed the driver if the accumulator pressure fell below 70 lbf/in².

8.4.2 Control system

The complete installation is shown schematically in Fig. 8.12. The Delco Products Division were responsible for the design of the automatic levelling system. Besides a new lighter levelling valve, they developed a system of solenoid valves to allow the passage of air to the levelling valves when the ignition was switched on or when a car door was opened. With closed doors and the ignition off the solenoid valves isolated the levelling valves. A second pair of solenoid controls were used to allow a restricted air flow to pass in and out of the air springs when the vehicle began to move. These solenoids operated only on demand from the levelling valves. Such action would be required if, for example, the vehicle had been parked on a steep hill or with one wheel in a pothole.

As in the earlier system designed for the Greyhound buses, a delay mechanism was built into the levelling valves. No air flow into or out of an air spring could take place under the action of a movement that had a frequency greater than 10 cycles per minute (0.167 Hz). This ensured that each air spring functioned as an individual unit under the action of normal bump and rebound movements. Three levelling valves were used, as in the bus installation, a single valve at the front and two at the rear.

References

[1] Sainsbury, J. H. (1958), 'Air suspension for road vehicles', *Proc. Inst. Mech. Engrs.*

[2] *Air Suspension and Servocontrolled Isolation Systems*, Shock and Vibration Handbook No. 3, McGraw-Hill, New York, chapter 33, p. 331.

9
Hydropneumatic Suspension

9.1 The Citroën tradition

Engineers at Citroën have always been encouraged to be bold
in their thinking, even though the commercial history of the
company suggests that boldness can sometimes lead to
financial instability. André Citroën was the first automobile
manufacturer to make front wheel drive cars in large-scale
production. All Citroëns since 1934 have been *traction avant*
and it is inconceivable that they will ever be otherwise. Even
though Citroën did fail financially in the 1930s, when an
injection of money from the Michelin Co. kept them in
business, the bold approach to engineering design was not
abandoned.

While Firestone and General Motors were wrestling with
the problems of pneumatic suspension in the 1950s, Citroën
forged ahead with their plans for a self-levelling suspension
that used a gas as the suspension medium and a liquid to
achieve the automatic levelling.

9.2 The Citroën suspension

Hydropneumatic suspension was first used in 1953 for the
rear suspension only, on the 6-cylinder Citröens. In 1955 it
was introduced on the DS model. The DS series of Citroëns
was replaced by the CX series, a remarkable vehicle, using
high-pressure hydraulic circuits to operate not only a
self-levelling suspension system, but the brake system, an
assisted steering gear and an automatic control of clutch and
gear change. In this book our interest is in the suspension
only, but the reader must appreciate that the provision of an
expensive hydraulic pump and its ancillary controls is a
shared service. For obvious reasons provision is made in the
main hydraulic circuit, in the shape of a 'priority valve', to

isolate the suspension system and maintain normal hydraulic supply and pressure to the brakes and steering in the event of serious failure in the suspension system.

9.2.1   The pressure circuit

The heart of the system is the hydraulic pump, and to extend the anthropomorphic analogy, we can see from Fig. 9.1 that the hydraulic circuit contains arteries to supply oil under pressure and veins or return pipes to return the oil to the pump inlet. From this point our analogy begins to break down. The Citroën engineers, being only human, found it necessary to add several control devices that appear clumsy when compared with the self-regulating devices found in the human heart.

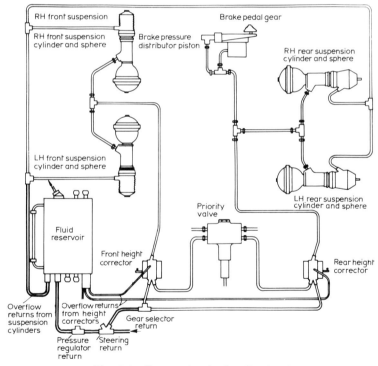

Fig. 9.1 Suspension hydraulic circuit.

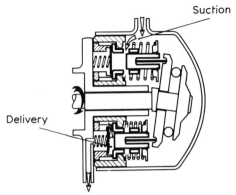

Fig. 9.2 Simplified cross section of swashplate multicylinder pump as used on CX model.

The Citroën pump on the larger models, the DS and CX series, is an engine-driven, 7-cylinder swashplate pump, shown in a simplified section in Fig. 9.2. It operates at half-engine speed. On the GS model it is a single cylinder unit. The pump delivery on the larger cars is 2.88 ml per revolution. Since September 1966 (December 1968 in the USA and Canada) a mineral oil, similar to engine-lubricating oil, has been used as the hydraulic fluid. A spring-controlled regulator is used to maintain a pressure in the main hydraulic

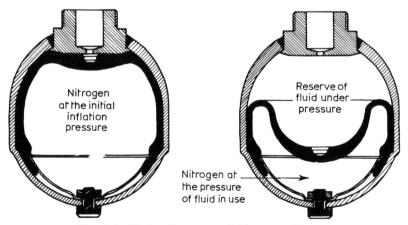

Fig. 9.3 Hydraulic system fluid accumulator.

circuit within the designed limits. The regulator is set to cut in at 140–147 bar (2030–2130 lbf/in²) and cut out at 162–167 bar (2350–2420 lbf/in²). These limits are wide enough to permit a reserve of fluid to be stored in a gas 'spring-loaded' accumulator. This reduces the number of starts and stops demanded from the pump on a journey. Cross sections of the accumulator are given in Fig. 9.3. The view on the left is of the vessel in the original condition with the zone below the

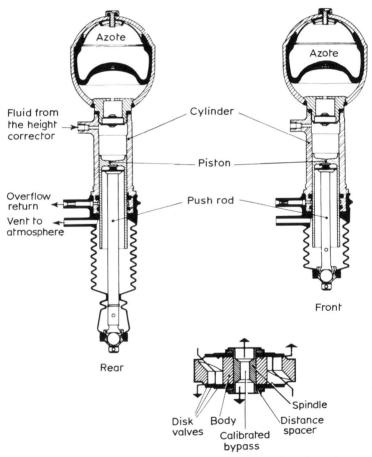

Fig. 9.4 Front and rear suspension cylinders with enlarged cross section of damper valve body.

synthetic rubber diaphragm at the initial charge pressure. The gas is dry nitrogen. On the right the accumulator is fully charged with hydraulic fluid at the regulator cutout pressure. The pump is supplied with fluid from the fluid reservoir which is not only a reservoir, but a filtration and degassing tank into which all fluid is returned from the various operating units, such as the suspension units, gear selector, etc.

9.2.2   The suspension circuit

The sprung mass is supported on four units of the type shown in cross section in Fig. 9.4. These are the type of unit used on the medium-sized Citroën, the GS model. The upper part of the 'sphere' is charged with nitrogen (*azote* in French); the lower part and the damper cylinder contains hydraulic fluid. The gas cylinder is obviously not spherical, having a cylindrical central zone. This profile is carefully designed by Citroën to give the desired increase in effective piston area as the nitrogen is compressed. This change in effective piston area is illustrated schematically in Fig. 9.5. In the upper diagram a wheel has risen in bump and the piston area is almost the full 'sphere' diameter. In the lower diagram the gas has expanded as the wheel falls into a pothole and the diaphragm, having forced fluid from the cylinder, has now a much-reduced effective piston area. This change in area with diaphragm movement should be compared with the use of a

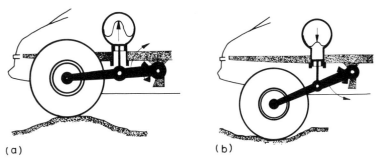

(a)   (b)

Fig. 9.5 (a) Compression of the gas spring under single-wheel bump. (b) Expansion of the gas spring under rebound.

tapered or curved pedestal to achieve a similar effect in the
Cadillac pneumatic suspension (see Fig. 8.8).

The movement of hydraulic fluid into and out of the lower
part of the 'sphere' is controlled by a damper plate in the
base. The damping action is double-acting. Small wheel
movements are absorbed by flow through the calibrated
bypass hole. Larger flows in both directions are controlled by
upper and lower disk valves. In this respect, the dampers
follow conventional damper practice.

9.2.3 Gas pressure

The load on each wheel is carried by the gas above the
diaphragm. The initial pressurization of each unit is related
directly to the weight carried. On the GS saloon, for example,
with a weight distribution of 62/38, each front suspension
unit is pressurized to 55 bar, the rear to 35 bar. In the
absence of wheel movement the pressures above and below
the diaphragms balance out to the same value. In the static
position each diaphragm will, therefore, be flexed to some
position similar to that illustrated in Fig. 9.4.

9.2.4 Height correction

Two height correction valves are used, one at the front and
one at the rear. These feed or extract fluid from the
suspension units to maintain a constant ground clearance.
Relative rotation of the anti-roll bar is used as an indication of
the correct ride height. The height corrector valve is shown in
section in Fig. 9.6. There are four pipe connections to the
valve. Three are shown in the upper cross section, these
being an inlet from the hydraulic pressure supply and a
connection to the reservoir at the top, an outlet to the
suspension units at the base. The lower view in Fig. 9.6 is a
half-section of the valve, but on a different plane passing
through the restricted passage (also called a 'dashpot' in the
Citroën literature) and the large-bore passage connecting
chamber D with chamber C. The action of the valve is
explained in the text included with Fig. 9.7.

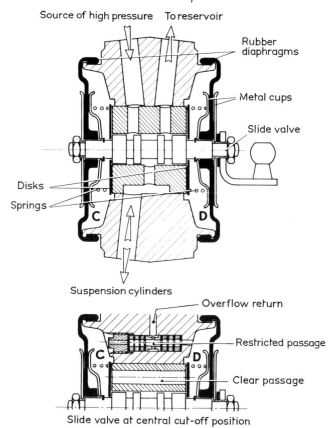

Source of high pressure   To reservoir

Rubber diaphragms

Metal cups

Slide valve

Disks

Springs

C   D

Suspension cylinders

Overflow return

Restricted passage

C   D

Clear passage

Slide valve at central cut-off position

Fig. 9.6 Height corrector: a distributor block (3-way tap) which, depending on the position of the slide valve connects the services (suspension cylinders) to the inlet (HP supply); connects the services (suspension cylinders) to the outlet (reservoir); isolates both the inlet and outlet from the services (slide valve central). The chambers C and D, sealed by rubber diaphragms (reinforced by metal cups), are full of hydraulic fluid which comes from the seepage past the slide valve; a seepage return takes the surplus fluid back to the reservoir. Chambers C and D are interconnected by a clear passage drilled in the sleeve of the slide valve, closed at each end by disk valves controlled by the movement of the slide valve. In the central position each disk is held against a face on the sleeve by a weak spring. A restricted passage inserted in the body of the corrector (dashpot) which limits the flow of fluid from C to D and back is connected to the overflow return to the reservoir.

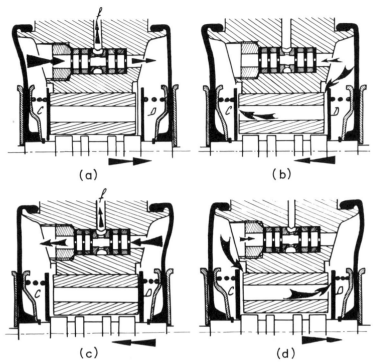

(a)

(b)

(c)

(d)

Fig. 9.7 Operation of the height corrector. (a), movement from *cutoff* to *exhaust* position: When the slide valve is moved, i.e. when it moves its position from cutoff, the disk valve in chamber C is held on its seating by a return spring, thus closing the clear passage. The disk valve in chamber D is lifted off its seating by the shoulder on the slide valve, thus opening the free passage. The fluid in chamber C is therefore obliged to pass through the dashpot which slows down the fluid movement, which in turn slows down the movement of the slide valve. Thus the slide valve will not move to the exhaust position unless there is a positive effort on it for a certain period of time. No correction occurs for rapid wheel movements; (b), movement from *exhaust* to *cutoff* position: When the slide valve is returned to the cutoff position, the fluid in chamber D can this time use the clear passage and return to chamber C, lifting the disk valve against its return spring. Return is therefore rapid. As soon as the slide valve returns to the cutoff position, the disk valve in chamber D closes the passage again, stopping the slide valve over-running the cutoff position and avoiding a second correction; (c), movement from cutoff to inlet position: When the slide valve is moved, the disk valve in chamber D is held against its seating by its spring, closing

9.2.5   The function of the anti-roll bar in height control

Rotation of the anti-roll bar is used to control the movement
of the slide valve. We must first consider how rotation of the
anti-roll bar occurs. During roll one arm of the anti-roll bar is
forced upwards, the other arm downwards. Since these
movements are equal and opposite the bar is placed under
torsion, but no actual rotation occurs relative to the
supporting bushes carried by the body. Simple single wheel
movement in bump or rebound will produce some torsion
and a little rotation. However, if both front wheels or both
rear wheels move upwards or downwards in unison, the
anti-roll bar arms on both sides move *in the same direction*.
This produces rotation of the anti-roll bar and is the basis of
the Citroën system of height control.

The complete control mechanism is shown in Fig. 9.8. The
height control rod is clamped to the anti-roll bar and passes,
parallel to the bar, to a lever behind the corrector valve. This
lever is pivoted at its lower extremity. Rotation of the anti-roll
bar moves the corrector valve slide inwards or outwards, as
described in Fig. 9.7. Rapid wheel movements, i.e. under the
action of single- or double-wheel bump or rebound, have no
effect on the corrector valve, since the movement of fluid
between chambers C and D is restricted by the 'restricted
passage' shown in Fig. 9.6. Height correction can only occur
after a delay of several seconds.

An over-riding manual control, located near the driver,
allows him to increase the ride height when travelling over a
rough terrain or to clear snowdrifts. It is also used when
changing a wheel.

---

the clear passage. The disk valve in chamber C is lifted off its seating
by the shoulder on the slide valve, thus opening the clear passage.
Liquid in chamber D therefore has to pass through the dashpot. As
for operation (a), there will be no movement of the slide valve until
a certain designed time period has occurred; (d), movement from
inlet to *cutoff* position: When the slide valve is returned to the cutoff
position, the fluid in chamber C can this time use the clear passage
and return to chamber D, lifting the disk valve off its seating. As in
operation (b), the return is rapid, As soon as the cutoff position is
reached, the disk valve in chamber C reseats. This stops any
over-run of the slide valve and prevents a second correction.

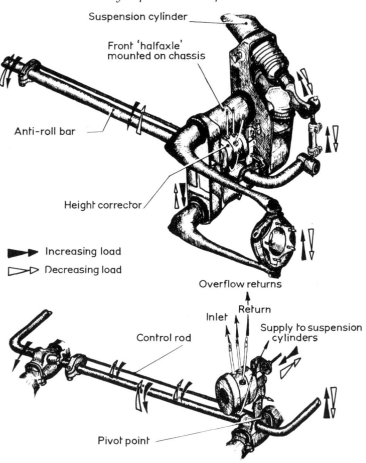

Suspension cylinder

Front 'halfaxle'
mounted on chassis

Anti-roll bar

Height corrector

▶ Increasing load
▷▷ Decreasing load

Overflow returns

Return
Inlet

Supply to suspension
cylinders

Control rod

Pivot point

Fig. 9.8 Activation of the height corrector valve by rotation of the anti-roll bar.

The Citroën control system is excellent, for a mechanical system, but there is little doubt that a much lighter and cheaper electronic control system will be seen in the not-too-distant future

9.2.6   The GS front suspension

On the GS model a double-wishbone layout is used for the front suspension with the spring unit acting on the upper

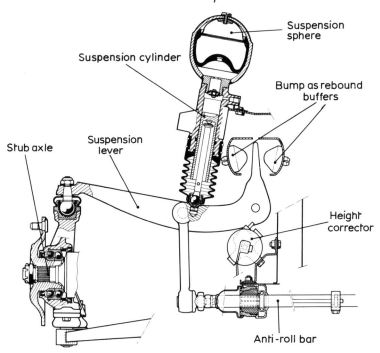

Fig. 9.9 Front suspension on GS model.

wishbone. The piston stroke of 70 mm becomes a total movement of 210 mm at the wheel. The mechanical advantage can be seen in Fig. 9.9. With a self-levelling system this total wheel movement of 210 mm is available even when the vehicle is fully laden. The spring rates for small wheel movements are exceptionally low. Measured at the wheel they are approximately 5 kN/m one-up rising to 6 kN/m fully laden. The natural frequencies given by these two rates are 0.65 Hz and 0.70 Hz respectively. These frequencies are nearly one-half the values given by coil springs on conventional vehicles. The spring rates in bump rise rapidly as the deflection increases. Not only does the gas pressure rise under compression, but the diaphragm area increases. In rebound the spring rates and frequencies are still very low for small wheel movements. In this direction, however, the gas pressure falls as the deflection increases and the diaphragm area also becomes much smaller near the limits of rebound movement. The flow characteristics of the damper valves also influence the effective spring rates. By providing stronger springs on the rebound damper valves, Citroën are able to provide an adequate control on wheel movement in rebound. Rubber buffers, indicated as item 1 in Fig. 9.9, act as bump and rebound stops.

9.2.7 The GS rear suspension

A trailing link rear suspension is used with the suspension unit set at an angle of $17°$ to the horizontal. The general layout can be seen in Fig. 9.10. The suspension unit is hidden by the subframe in this view; only the top of the cylinder is visible. The details of the suspension cylinder and sphere can be seen in Fig. 9.11. For small wheel movements the frequency in bump and rebound at the rear is about 15% higher than that at the front.

9.2.8 The anti-roll bars

The rear anti-roll bar is 18 mm diameter, the front 21 mm. Since torsional stiffness varies as the fourth power of the diameter, the front anti-roll bar rate is 85% greater than the rear. Even though the centroid of the sprung mass has a strong bias towards the front, there will still be a greater roll

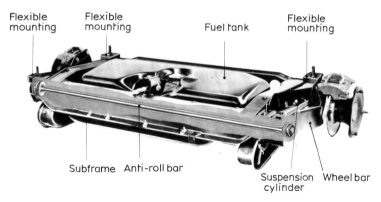

Fig. 9.10 Rear suspension installation on GS model, showing anti-roll bar.

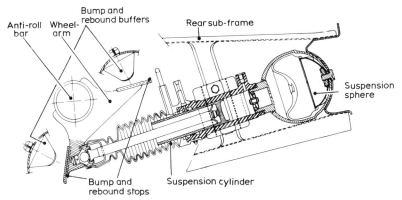

Bump and
rebound buffers

Anti-roll   Wheel-
bar   arm

Rear sub-frame

Suspension
sphere

Bump and
rebound stops

Suspension cylinder

Fig. 9.11 Rear suspension cylinder on GS model and its installation
details.

resistance at the front and a slight tendency to roll-oversteer
when driven near the limit in a corner.

The Cadillac engineers did not use anti-rolls bars in their
self-levelling system, and with a delay mechanism built into
the height control mechanism there would be a tendency to
roll excessively when making rapid changes in direction. This
problem does not arise with the Citroën design.

9.2.9 Ride-height adjustment

I have seen how useful this device can be. I lived at one time
within a kilometre of the west shoreline of Lake Michigan,
seeing my doctor neighbour in his DS model Citroën charge
quite successfully through formidable snowdrifts. On such
mornings he jacked up the Citroën to its full 254 mm (10 in)
ground clearance. His most reliable Volkswagen Beetle was
used only after the city snowploughs had cleared the streets.

9.2.10 Cornering power

Only in one sense does the Citroën appear old-fashioned.
Trailing link suspension is used at the rear and this, we
know, gives undesirable camber changes in roll. At the front,
although the double-wishbones are angled inwards towards
the car centre line, they are almost of equal length and will

give an undesirable positive camber to the outer wheel under roll. The Citroën, therefore, does not appear to achieve the full potential of the cornering force available in the tyres. Other suspension geometries are possible and offer greater cornering potential. This will be seen when we examine the latest European application of hydropneumatic in the S-class Mercedes-Benz 450 SEL later in the chapter.

9.2.11   Anti-squat

It would be very odd indeed if Citroën introduced all this complex hydraulics into the suspension design in order to provide a constant ride height and then completely ignored the longitudinal dips and dives induced by braking and acceleration.

Anti-dive geometry in general was discussed in Chapter 4. The centroid in the Citroën is well forward, with a one-up weight distribution of 62/38 and a full-load distribution of

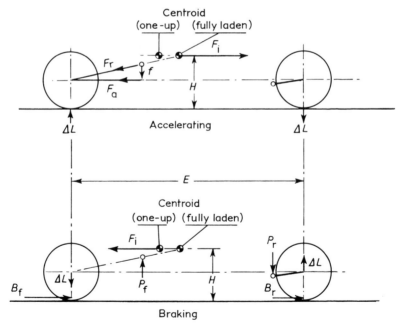

Fig. 9.12   Anti-squat and anti-dive geometry on GS model.

54/46. The force diagram in Fig. 9.12 is a compromise based on a weight distribution of 58/42.

To achieve perfect anti-squat geometry, the decreased loads on the front footprints and the increased loads at the rear, as shown in the upper diagram in Fig. 9.12, must be balanced by a vertical downthrust on the suspension arms at the front. The tractive force F from the front wheels produces an acceleration α. Approximately 10% of this tractive force is expended in accelerating the unsprung mass and about 90% in accelerating the sprung mass M_s:

$$0.9F = F_i = M_s\alpha.$$

F_i is the inertia force acting on the sprung mass. This tends to rotate the sprung mass about its suspension anchorages. This decreases the load on the front wheels by ΔL and increases the rear wheel loads by the same amount:

$$\Delta L = F_i \frac{H}{E} = M_s\alpha \frac{H}{E}.$$

The effective arms of the two moments H and E can be seen in Fig. 9.12.

To balance this tendency for the body to rotate during acceleration, the front wishbones have been angled upwards on the Citroën GS at an effective leading arm angle of 12 degrees. Taking $F_a = F_i = 0.9F$, where F is the total tractive force, $f = F_a \tan 12°$.

From Fig. 9.12 it is seen that perfect anti-squat geometry is achieved when the resultant force F_r is angled towards the centroid. The diagram shows that almost perfect anti-squat balance is given when fully laden. With a one-up load about 80% anti-squat is given.

9.2.12 Anti-dive

With a trailing arm suspension at the rear and a wishbone system at the front, angled to give an effective leading arm action, the Citroën GS has an ideal geometry for anti-dive correction. As shown in the lower diagram in Fig. 9.12, the values of ΔL now become an increase in load at the front and a decrease at the rear. The braking forces applied to the footprints B_f at the front and B_r at the rear, produce reactions at the suspension points P_f upwards at the front and P_r downwards at the rear. These forces, with careful design, will balance the couple $\Delta L \times E$.

The Citroën GS has a brake pad area at the front of 146 cm^2 and at the rear of 68 cm^2. This suggests a front/rear braking effort of 68/32. The situation is more complicated, however, since the front and rear brake cylinders do not operate at the same pressures or even at constant pressures. The front brakes always operate at normal full-circuit hydraulic pressure (*circa* 160 bar). The rear brakes operate at rear suspension pressure, which is nominally 35 bar. When fully laden the rear suspension pressure increases to as much as 65 bar, but this is still well below the operating pressure at the front. It is obvious that the front braking bias is so high under all load conditions that the front wheels will always be the first to lock. This is an established prerequisite for safe braking in the wet. The lower diagram in Fig. 9.12 is given purely to show the principles of anti-dive geometry, using leading and trailing arms. We do not have sufficient data to put concrete values to the Citroën design.

9.3 The Mercedes-Benz hydropneumatic suspension

The engineer has always been in control at Daimler-Benz. The German for 'stylist' is no doubt a very long word, but his influence is relatively short. Market forces are not ignored, but a new body on a Mercedes-Benz is normally only introduced as outer covering for a new advanced design of car. The engineering comes first, the styling second.

Having abandoned their long battle with the swing axle rear suspension, Daimler-Benz adopted the semi-trailing arm rear suspension and refined it. In conjunction with a well-developed double-wishbone system at the front, using unequal-length wishbones suitably angled to give anti-dive geometry, the mid-1970s range of Mercedes-Benz cars seemed to represent the ultimate in conventional suspensions. When tested by *Motor* in 1974 the 450 SEL was pronounced as 'simply magnificent'. Even while the *Motor* staff were revelling in the 'uncannily roll-free, sure-footed handling', the research-and-development engineers at Stuttgart were working on a hydropneumatic suspension.

9.3.1   The 450 SEL 6.9

The 1979 450 SEL is powered by a 6.9 litre V8 engine of 210 kW (286 h.p. DIN), with K-Jetronic petrol injection. The most interesting feature, however, is the application of gas suspension with hydraulic level control to the well-established mechanical suspension components from the earlier S-series Mercedes-Benz chassis.

Interest in the Experimental Department in hydropneumatic suspension began in the 1950s, obviously inspired by the Citroën DS 18. Daimler-Benz made a serious study at the time of the influence on roll and pitching of the use of gas springs. The first application of this work was a hydropneumatic shock-absorbing strut used as a compensating spring in conjunction with the low-pivot, swing axle suspension used on later models of the 300 SL sports car. The next development was a self-levelling rear suspension using hydropneumatic struts with pressure supplied by an engine-driven pump. This, in turn, progressed to a fully automatic hydropneumatic suspension à la Citroën. This is available on the S-class top model, the 450 SEL 6.9.

There are improvements on the Citroën system. For example, suspension height can change when the engine is not running. This is because the central reservoir is charged to a higher pressure than that required for normal operation. This high-pressure reservoir is particularly valuable when changing a flat tyre. The pressure line to any particular suspension unit can be isolated during a wheel change then repressurized from the reservoir afterwards. If the pressure supply fails by pump failure or other cause, both axle circuits can be isolated to complete the journey. Another improvement is the basic suspension geometry which, unlike that on the Citroën, is designed to maintain near-verticality of the outer wheels when cornering.

The layout of the hydraulic system is shown in Fig. 9.13 and the placing of the suspension components in Fig. 9.14. One clever modification of the Citroën system is the use of a

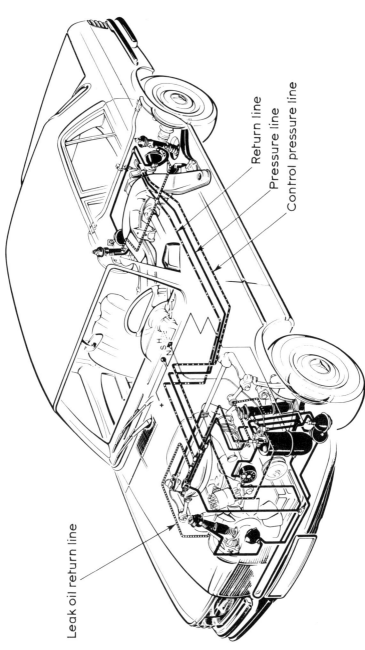

Leak oil return line

Return line
Pressure line
Control pressure line

Fig. 9.13 Hydraulic circuit for hydropneumatic suspension on Mercedes-Benz 450 SEL 6.9.

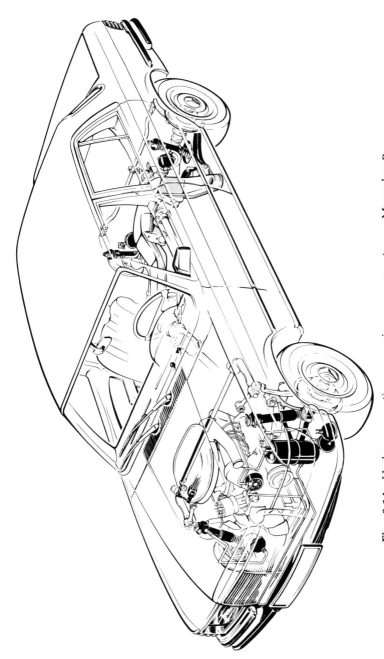

Fig. 9.14 Hydropneumatic suspension components on Mercedes-Benz.

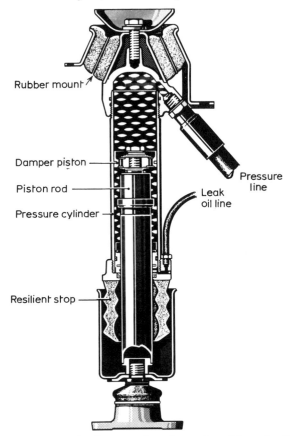

Rubber mount

Damper piston

Piston rod

Pressure cylinder

Pressure line

Leak oil line

Resilient stop

Fig. 9.15   Front wheel suspension element on Mercedes-Benz.

large-bore connecting pipe between the hydraulic suspension cylinder and the gas 'sphere'. This gives the designer more freedom in the disposition of the components. Fig. 9.16 shows how the gas reservoir (the sphere) can be fitted neatly into the side of the rear wheel arch. The Citroën engineers solve the difficult problem of intrusion into the passenger space by placing the cylinder/sphere unit in a near-horizontal position. Fig. 9.15 shows a cross section of the Mercedes-Benz front suspension cylinder, and Fig. 9.16 shows the rear suspension layout. It will be seen that a

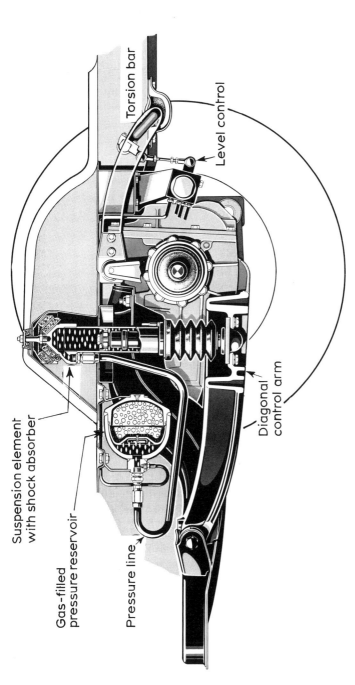

Suspension element
with shock absorber

Torsion bar

Level control

Gas-filled
pressure reservoir

Diagonal
control arm

Pressure line

Fig. 9.16 Rear wheel suspension layout on Mercedes-Benz: 1, pressure line; 2, gas-filled pressure reservoir; 3, suspension element with shock absorber; 4, torsion bar; 5, level control; 6, diagonal control arm.

variant of the Citroën level control method is used. Angular movement of the centre point of the anti-roll bar (see Fig. 9.14 for the position of level control clamp) is used to actuate the level controller (item 5 in Fig. 9.16). The detail design of the suspension cylinder can be seen in Fig. 9.15. The damper valves are carried in the piston, not in the base of the gas reservoir as in the Citroën. Any leakage past the piston rod seals is returned to the main reservoir via the special return lines. The hydraulic supply reservoir is the large cylindrical vessel which can be seen in Fig. 9.13 situated in front of the front suspension cylinder on the right. The gas vessel associated with this reservoir can be seen adjacent to the suspension 'sphere' on the right of the drawing.

Anti-dive and anti-squat had already been built into the 450 SEL suspension before the adoption of hydropneumatic suspension. Anti-roll provision, however, had to be supplemented. With gas suspension bump and rebound rates very much lower than those used with coil suspension, it was necessary to provide much stiffer anti-roll bars to achieve the same roll resistance.

10

Interconnected and No-roll Suspensions

10.1 Interconnected front/rear suspension

The concept of interconnecting the front and rear suspensions to reduce the period in pitch is an old one. A mechanical interconnect was developed for the little Citroën 2CV and this, despite an obvious lack of roll stiffness, gave a very good ride over rough terrain for such a tiny vehicle.

Let us consider as a first step a simple system where both front and rear suspensions share the same spring, i.e. one spring per side as shown in Fig. 10.1. With a perfectly symmetrical system the sprung mass could not oscillate in pitch, since a downforce at A would deflect the front spring connection a distance x, and this would be balanced by an upforce at the rear of identical value which would deflect the rear spring connection an equal distance x *in the same direction*. Spring deflection would therefore only occur in bump and rebound, not in pitch. Before we rush off to the Patent Office, let us consider what would happen to such a vehicle on the road. As soon as we started to move from a standstill, even at a negligible rate of acceleration, the inertia force of the sprung

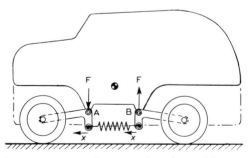

Fig. 10.1 An interconnected suspension, but with no resistance to pitch.

mass would rock the body backwards, pushing downwards at B and pulling upwards at A. There would be no spring force to act against this inertia force and the body would tip backwards hard against the limit stops. Light braking would produce the opposite reaction, since the system is completely unstable in pitch.

10.2   The Citroën 2CV

Additional springs *without interconnection* are seen to be essential to achieve a practical interconnected suspension. This can be seen on the Citroën 2CV, shown schematically in Fig. 10.2 in which the two main springs S_f and S_r are enclosed in a 'floating' cylinder, which is mounted between auxiliary springs s_f and s_r with abutments against the sprung mass. In the later examples of the 2CV these auxiliary springs are made of rubber.

Considering a simple system with a weight distribution of 50/50 and identical springs rates front and rear ($S_f = S_r$ and $s_f = s_r$), we have a system in which bump and rebound is controlled by two springs S and s working in series and pitch is controlled by two springs S and s working in opposition. When two springs of rates S and s work in series, they act as if replaced by a single equivalent spring of rate S_e where:

$$S_e = \frac{S \times s}{S + s}\ .$$

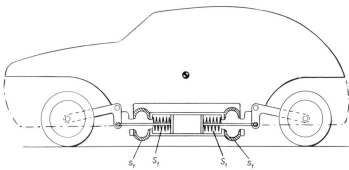

Fig. 10.2   An interconnected suspension as used on 2CV Citroën.

If, for example, we make $s = 0.7S$:

$$S_e = \frac{S \times 0.7S}{S + 0.7S}$$

$$= 0.412S.$$

S_p, the rate in pitch $= S - s$
$$= 0.3S.$$

This would give a natural frequency in pitch (since natural frequencies vary as the square root of spring rates) of about 85% of the natural frequency in bump and rebound. The above analysis is an oversimplification, but serves to illustrate the Citroën principle.

10.3 The Moulton hydrolastic suspension

A mechanism using hydraulic interconnection of front and rear suspensions was filed with the British Patent Office in

Fig. 10.3 Idealized behaviour of the MG 1100, with interconnected suspension units, when traversing a bump in the road.

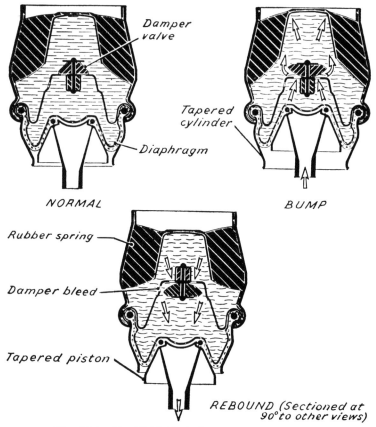

Damper valve

Tapered cylinder

Diaphragm

NORMAL

BUMP

Rubber spring

Damper bleed

Tapered piston

REBOUND (Sectioned at 90° to other views)

Fig. 10.4   The BMC Hydrolastic suspension unit.

1955 by Alex Moulton and tried out experimentally in the same year on a prototype Alvis that had been designed by Alec Issigonis.

The first production application of Hydrolastic suspension, the name adopted by Moulton Developments Ltd, was to the BMC 1100. An idealized diagram of the behaviour of *hydrolastic* suspension on a variant of the original ADO 17 design, the MG 1100, is shown in Fig. 10.3. The suspension unit is shown in section in Fig. 10.4. It consists of a tapered sheet-metal casing, fixed to the body frame, which encloses in the upper section a conical rubber spring. At the base is a sheet-metal partition. The outer rim of this and the lower extremity of the casing are swagged around the reinforced rim of the moulded rubber/nylon reinforced diaphragm. Two rubber one-way valves in the centre of the partition permit flow in both directions. These two valves, together with the small bleed hole, replace the conventional damper system. The fluid on both sides of the partition is a mixture of water, alcohol and anti-corrosion agent. The piston rod which is located in the centre of the diaphragm is connected to the suspension links and moves up and down with wheel movements.

10.3.1 Interconnection

The Hydrolastic spring elements on the same side of the car are interconnected by small-bore piping. The pipe connection cannot be seen in Fig. 10.4 but is made at a high point on the upper surface. Fig. 10.3 illustrates how single wheel movements were controlled by both front and rear springs on the same side working *in series*. With very large-bore piping this would have effectively halved the spring rate in comparison with the use of isolated units. In practice the use of small-bore piping modified the true series operation to an intermediate effective spring rate. The damper design also modified the overall behaviour. Under roll, however, the two springs on the same side worked in parallel and, as we know, springs working *in parallel* are additive.

The Moulton Co. have a long experience of the design of rubber suspension units and they were able to provide a rubber doughnutlike spring (see Fig. 10.3) which had been carefully designed to give a rising rate suspension. With very large-bore pipes the only resistance to movement in pitch would be the resistance to flow designed into the damper valves. With small-bore pipes the pitch frequency must increase for larger movements. From experience, I can state that pitch control on this vehicle was very good and for such a small car the ride was excellent. Today some small cars with conventional springing have an inferior ride.

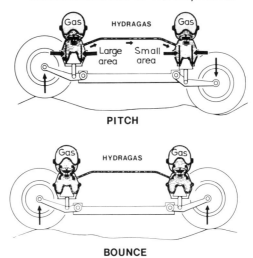

PITCH

BOUNCE

Fig. 10.5   The Hydragas principle.

10.4   The Moulton Hydragas suspension

The success of the Citroën hydropneumatic suspension no doubt inspired the Moulton Co. to turn to gas as the suspension medium; also to use a tapered pedestal or piston to give a variable effective diaphragm area, the technique originally used by Firestone.

The Hydragas principle is shown schematically in Fig. 10.5, and the detail design of a Hydragas spring unit is shown in Fig. 10.6. It will be seen that the effective piston area acting on the diaphragm increases with increase of wheel movement in bump and decreases under rebound. Nitrogen gas under pressure and compressed by a flexible diaphragm acts as the spring. Even in the position of the damper valve, the Moulton gas spring is almost identical to the one in the Citroën system. In the Hydrolastic suspension system the interconnection between front and rear was made from the liquid compartment *above* the damper valves. In the Hydragas system the connecting pipes are *below*. This important difference means that liquid flow between the front and rear units does not pass through the damper valves in the latest

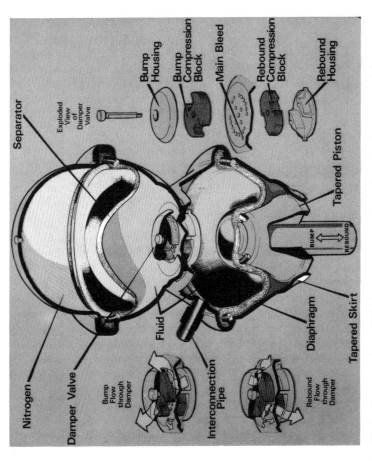

Fig. 10.6 Cross section of Hydragas suspension unit, showing parts of damper valve.

system. The main resistance to pitch is given by the change in diaphragm area by the piston taper, as shown in Fig. 10.5. Since one diaphragm increases in effective area as the other decreases the system behaves in pitch like a conventionally sprung car, but with no conventional damping. Some damping is given by the long interconnecting pipes.

Rates. The makeup of the bounce, roll and pitch rates, using the terms used by Moulton Developments Ltd, is as follows:

*Hydraulics** – that due to the compression of the gas acting through the liquid (at the same pressure) upon the diaphragm.

Taper – that due to the pressure of the liquid acting upon the changing area of the diaphragm, as it is actuated by the suspension arm.

Parasitic – that due to rubber bushings in the complete suspension system.

Drop angle – that due to change of leverage with stroke.

Essentially in the Hydragas system the pitch rate is much lower than in bounce; the bounce rate and the roll rate are identical since no anti-roll bars are used. As shown in Fig. 10.7, the pitch rate is made up of *taper*, *parasitic* and *drop*

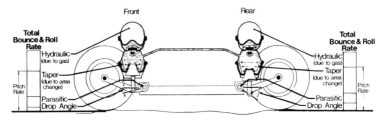

Fig. 10.7 Makeup of bounce, roll and pitch rates in Hydragas system.

---

*   Why this is not called 'Gas' or 'Pneumatic' rate is not clear.

angle. The bounce and roll rates include these three plus the *hydraulic* rate.

On a typical Hydragas suspension such as the Austin Allegro the pitch rate would be 13 kN/m, the front bounce and roll rate, 20 kN/m and the rear bounce and roll rate, 18 kN/m. On the 1750 HL Allegro, with a fully laden sprung mass of about 1050 kg and a front/rear weight distribution of 63/37, the front period in bounce is approximately 1.2 Hz and the rear 1.5 Hz. The pitch frequency is about 1.0 Hz, a very low value and well removed from the bounce frequencies. I myself have used an Allegro for personal transport for several months and found the ride to be as good as that given by many larger cars. The low pitch frequency is no doubt responsible for the favourable comments on the good ride given in the back seat.

10.5 No-roll suspensions

10.5.1 Roll, no-roll or banking

Before wasting time and money on making a product, we should always establish that a need really does exist. The Ford Co. no doubt still remembers the Edsel. There was nothing much wrong with the Edsel, but the American public was not in the market for yet another big car when they introduced it!

Very little market research seems to have been carried out to discover how much the typical carowner will pay for a car with a better ride, or a reduced tendency to roll when cornering. Even the men who design the cars do not seem to be of one mind when the question of roll is considered. French cars, for example, usually roll more than British cars. What, indeed, is the ultimate suspension behaviour when cornering? Should there be a limited degree of roll, no roll, or even what is usually called 'banking', i.e. a reverse body roll to counteract the effects of centrifugal force on the occupants?

10.5.2 The banking car

The idea of banking the body, i.e. leaning inwards, when cornering has appealed to many inventors. It appealed especially to the designers of racing motor cycles fitted with sidecars. A banking sidecar seemed a more sensible approach than one where the sidecar passenger was hanging perilously from one side or the other. Even so the idea of the banking sidecar eventually died.

Against this failure we can see a successful application in British Rail's high-speed train, which has a mechanism to make the carriages bank up to an angle of $9°$ when negotiating curves. This is simply to reduce the side forces on the passengers. A similar mechanism could also be used on an automobile.

In 1961, I drove a Chevrolet Impala sedan that had been converted by Mr J. Kolbe of Menomenee Falls, Wisconsin, USA to bank when cornering. Measurements taken from a photograph show the stock Impala with a body roll of about $6°$ on a particular turn. At the same speed on the same curve the Kolbe-converted Impala banked to an angle of about $2°30'$. The Kolbe Chevrolet was a very restful car to drive in, travelling along winding country roads, since the full-width front seat of the Impala gave very little lateral support to driver or passengers. In Suffolk, where I now live, a banking car would be a welcome innovation. In the USA it was difficult to find suitable winding roads on which to test out the Kolbe car. In general American roads travel in a straight line for mile after mile. Eventually, a $90°$ turn is made on to another long straight road. Little time is lost if the driver takes these $90°$ turns at a fairly low speed. Sad to relate, but Mr Kolbe invented the right car in the wrong country. The need must always come before the invention.

10.5.3 Practical no-roll suspensions

The advantages of no-roll suspensions have been enhanced in recent years by the increasing need to keep the wheels vertical in a corner. This applies more to sports cars and racing cars. For the typical tyres used on family saloons there is little reduction in cornering power until positive camber angles of $3–4°$ are exceeded.

Of course, it is possible to design front and rear suspensions with roll centres at the same height as the body-centroid centre line. With most vehicles this would require a very high roll centre at both ends and high roll centres result in large camber changes in single wheel bump and rebound, if we are to confine our design to conventional practice. This was clearly demonstrated in Fig. 5.15 in Chapter 5. With less popular suspensions such as the Morgan and the older Porsche parallel trailing arms there is no choice, since the roll centres are at ground level. The de Dion system gives a high roll centre, and it was rumoured several years ago that Team Lotus had considered the possibility of using a de Dion axle at both ends.

All Formula 1 and many Formula 2 cars in 1979 were what are popularly called 'ground-effect' cars. As was discussed in Chapter 6 these cars are designed to create a Venturi-effect underneath the car body to produce a surprising downforce thus increasing the available cornering forces in a high-speed corner by as much as 40% It is essential in such a system to maintain a good air seal between the sides of the body and the road surface. Sliding skirts have been developed for this purpose, but two critical factors threaten their viability. The first is the condition of the track surface, the second is the extent of the body roll. Modern billiard-table surfaces on Formula 1 circuits have made it possible for relatively short skirts to be used, moving up and down in the slides under bump and rebound and roll. Obviously, any tendency to roll

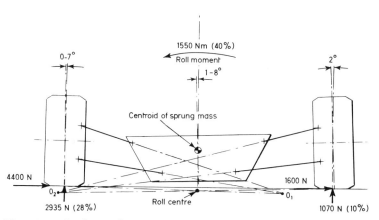

Fig. 10.8 Analysis of tyre cornering forces and roll moment at front of Lotus 79.

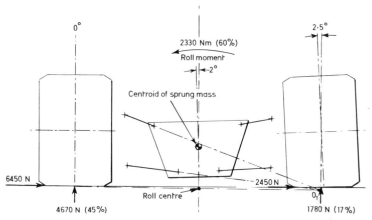

Fig. 10.9 Analysis of tyre cornering forces and roll moment at rear of Lotus 79.

on corners as much as typical racing cars of the early 1970s did calls for the provision of very cumbersome deep slide mechanisms.

Modern low-profile racing cars have placed a tight limit on roll for almost a decade. Even so, in 1978 when the Lotus 79 first appeared with sliding skirts, there were still a few cars that rolled as much as 3° when cornering near the limit, although the majority only reached about 2°, the limit for a viable ground effect car. Figs. 10.8 and 10.9 are based on measurements reported by *Motor* (March 3rd, 1979) in a 150 m.p.h. (67 m/s) corner and show the Lotus 79 rolling at a value very close to this limit.

The disposition of the main masses, the engine, transmission, radiators, fuel tanks and driver on a GP car are as close to the ground as the designer can put them. In the case of the typical 1979 Formula 1 car, the sprung mass centroid centre line slopes slightly upwards from about 150 mm above ground level at the front to about 180 mm at the rear. Since the roll centre is very close to ground level, the roll moment in a corner taken near the limit is very high. Extremely stiff springs are required to resist this moment. Even though progressive rate springs are used, the ride is so hard that young men in excellent physical condition feel battered at the end of a race.

Figs. 10.8 and 10.9 show that the inner tyres of an F1 only contribute about 30% of the total cornering force in a 150 m.p.h. corner. If then we could design a racing car with a 'no-roll' suspension in which load transfer did not take place, we could make more efficient use of the inner tyres and corner at even higher speeds. What is of equal importance (until they ban the ground-effect car!) is that a 'no-roll' suspension would assist the designer in the design of a ground-effect body.

10.5.4 The Trebron suspension

Norbert Hamy is a Canadian architect who invented the Trebron double-roll centre suspension more than a decade ago and has progressed, with the inevitable blind-alley developments and feasibility studies, to the latest Trebron Concordia, built as a prototype sports car with the collaboration of the Faculty of Engineering at Concordia University.

The earliest feasibility study was carried out by Harry Ferguson Research Ltd, resulting in the conversion of a Ford Escort by Broadspeed Engineering for Group 5 racing. Unfortunately, new racing regulations outlawed suspension changes in this group before this conversion could be properly sorted out on the race track but the initial impressions were very promising.

Detail drawings of the Concordia project have not been made available, but an earlier design for a Trebron Formula 1 chassis is given in Figs. 10.10 and 10.11. In this case the designer has concentrated on maintaining verticality of both outer and inner wheels. A camber angle of $0°-0°5'$ is no mean achievement at a cornering acceleration of $1.4g$. The secret ingredient of the Trebron system is the provision of a separate bulkhead with a roll centre RC_2 high in the air. Under the action of centrifugal acceleration the bulkhead rolls inwards, i.e. it banks. The roll centre for the suspension system RC_1 is at ground level. The banking action of the bulkhead produces a rocking action in the rocking links by which the upper wishbones are attached to the bulkhead. In the case of Fig. 10.11 the bulkhead has not only rolled inwards by $3°30'$, but shifted bodily sideways under

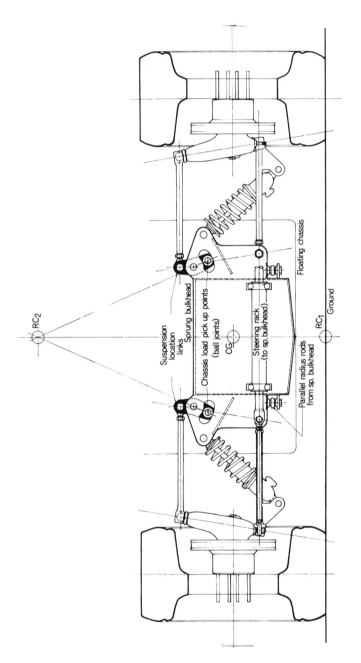

Fig. 10.10 Trebron double roll-centre suspension applied to racing car – static condition.

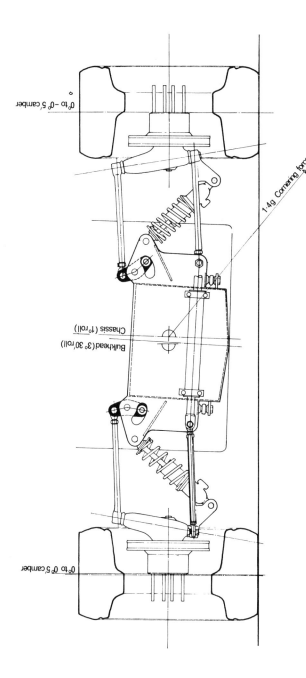

Fig. 10.11 Trebron DRC racing chassis under cornering acceleration of 1.4g. The suspension bulkhead swings about the upper roll centre (RC$_2$) and the wheels remain upright.

centrifugal force by an appreciable amount relative to the body (or chassis) centre line. The Trebron system shows great promise and might eventually be the answer to the contemporary racing car designer's most pressing problem.

10.6 The fully stabilized or 'active' system

Time and again, we have used the word 'compromise' in this introduction to the technology of automotive suspensions. There are fundamental limitations to conventional suspension systems. One basic limitation is that the static deflection varies as the inverse square of the natural frequency of the spring. Attempts have been made to overcome this problem by using rising rate springs, but this is only a palliative, and many research workers in the field have been studying the application of 'closed-loop' control systems to automotive suspensions and there have already been several experimental applications.

10.6.1 Active suspensions

A system is *active* as opposed to the conventional *passive* system when it uses a fast-acting, closed-loop control system of the type employed in aeronautics for many years. Such a system uses hydraulic rams, servovalves and sensors. A closed-loop control system requires the interconnection of these elements to feed back signals from the hydraulic rams by means of the sensors to activate the control device, which in turn feeds signals back to the rams. The number of sensors and the complexity of the control device depends upon the number of variables we find it necessary to control. Studies that have been made of the potential advantages of an active suspension system suggest the following:

  (a) a low natural frequency;
  (b) low dynamic and static deflection;
  (c) high speed response;
  (d) no change in frequency or response with change in load.

The disadvantages are the obvious ones of increased complexity and increased cost.

10.6.2   Automotive products no-roll suspension

Automotive Products Ltd have developed an active suspension system that is fully stabilized, i.e. pitch and roll are virtually eliminated and vertical accelerations are considerably reduced when compared with conventionally sprung cars. The system is still under development at their Leamington Spa research establishment, but a prototype form has been tested on a Rover 3500.

In the AP system the levelling valve no longer has a time delay built into it. Every effort is made to eliminate this time delay. If one stands on the front bumper of a Citroën CX model (with the engine running), the front end dips, then over a period of about two seconds, slowly rises to the controlled height. If one stands on the front bumper of the Rover 3500 equipped with the AP suspension system, nothing appears to happen since the levelling valve reaction time is a small fraction of a second. A line drawing of the AP system is given in Fig. 10.12. Power to correct ride height is provided by a pump capable of a very high oil flow at a pressure of 138–170 bar (2000–2500 lbf/in^2). It is a much larger pump than that used by Citroën. Full pump output, which would only be required occasionally as when cornering at the limit, is 10 kW (13.4 h.p.).

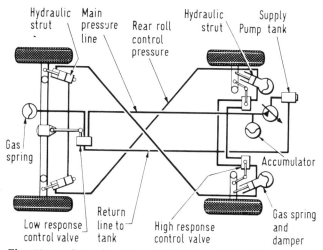

Fig. 10.12   Line drawing of AP fully stabilized suspension.

As in the Citroën system, each wheel is provided with a gas spring and damper strut which is connected to a suspension arm. The level of the car body under the action of roll or pitch is corrected by very rapid additions or extractions of hydraulic fluid from the space between the gas spring and the damper piston. The signal to add or extract fluid is given by a three-port valve. What the maker calls a 'pendulous mass' is used to sense any change in level (see Fig. 10.13). This mass is supported on a spring with a small hydraulic damper in parallel. For any application the mass-spring-damper unit must be specially tailored to match the suspension system. In effect, it is a replica of the car's suspension system in miniature. Thus under single wheel bump the action is as follows: upward movement of the suspension arm compresses the gas spring and creates an upward force to lift the body. The pivot of the offset pendulum also moves upwards, since it is attached to the body. If the pendulum did not move, the spool of the three-port valve would move to the left and extract fluid from the suspension leg. The pendulum-spring-damper unit, having been designed to behave as a replica of the suspension system, will produce an upward movement of the pendulous mass at an identical velocity to that of the body.

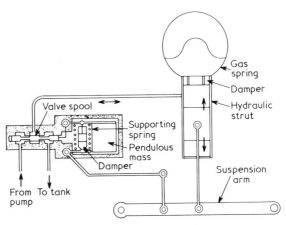

Control valve and suspension

Fig. 10.13 One suspension leg of the AP system, showing schematically the hydraulic control system.

The net effect is that the valve spool will not move in bump or rebound of a single wheel. The pendulum-spring-damper unit therefore acts as a filter for transient signals from a single unsprung mass, i.e. one wheel. Movement in roll, in brake dive or in acceleration squat – unless they are very transient as one might produce by a slight wiggle of the steering wheel – are usually at a lower frequency than single wheel movements. The valve spool reacts to these lower-frequency movements and fluid is either pumped into the appropriate struts or extracted from them at the combined dictates of the valves. In the experimental Rover two control valves are used at the front and a single control valve, with a lower response time, at the rear. The two front valves control the body level in pitch and roll, the rear controls only pitch. All three valves also maintain a constant body height irrespective of load, as in the Citroën and Mercedes hydropneumatic suspensions.

Early experimentation on the Rover installation showed that diagonal connection of the rear struts with the front gave good control of roll, but the rear struts had to be correctly sized relative to the front to give the correct balance between the front and rear roll couples. When the correct balance of strut sizes had been achieved, the Rover could negotiate bends and chicanes with precision and negligible roll. In fact the only observable roll was produced by tyre compression. When tested by the staff of *Motor*, the AP-stabilized Rover could negotiate a test chicane at 54.6 m.p.h. The limit for a standard Rover 3500 was 48.5 m.p.h.

The AP system would not be cheap. The pump alone would cost about £100, if produced in modest quantities. If one allows a total additional cost of £250, this would be a negligible increase on a car costing £15 000–£30 000, such as a Jaguar, BMW or Mercedes. In the price-conscious mass market it would not sell as well, unless perhaps one also included with the package a chromed embellishment in a prominent position on the body to tell the Jones's that the car was 'float-ride stabilized'.

References

[1] Moulton, A. E. and Best, A. (1979), 'From Hydrolastic to Hydragas', *Proc. Inst. Mech. Engrs*, vol. 193, no. 9.

[2] Moulton, A. E. and Best, A. (1979), 'Hydragas suspension', SAE paper 790374.
[3] Hegel, R. (1973), 'Vehicle attitude control methods', SAE paper 730166.
[4] Sutton, H. B. (1979), 'Synthesis and development of an experimental active suspension', *Automobile Engineer*, vol. 4, no. 5 (October–November).

11
A Small FWD Saloon Car: Ford Fiesta S

11.1 Front wheel drive

Today everyone seems to accept FWD, usually with a transverse engine and gearbox, as a *sine qua non* in the smallest class of four-seater family saloon. No other design gives as much space for the occupants and the luggage. FWD is effective, inexpensive and reliable. It was not always so. The first FWD car that I owned was a 1936 BSA Scout with front drive shaft couplings that produced unpleasant alternating reactions on the steering gear in a tight turn. Some drivers chose to ignore this kickback at the steering wheel; it is even possible that some regarded it as indication that they were cornering at too high a speed. To an engineer it was an indication of a variation in drive torque, something that would soon produce excessive wear. It was no consolation to me when my fears were proved right!

11.2 The constant-velocity universal joint

In this book we are not concerned with the transmission of power to the road wheels, but the rapid expansion of FWD automobiles since the end of the Second World War owes so much to the invention of the *constant-velocity joint* that we must consider in some detail the mechanics of this device.

Jerome Cardan, a sixteenth-century physicist, invented the first mechanical joint for transmitting torque from one shaft to another set at an angular displacement to the first. The first practical universal joint, however, was devised by Robert Hooke about 300 years ago and this well-known joint is illustrated in Fig. 11.1. This very simple joint, in many mechanical guises, was the basis of all mechanical U/J for

Fig. 11.1 Double Hooke's joint.

about 250 years. The alternative was the *flexible coupling* in which a leather, fabric or rubber disk was used to transmit rotary motion through a very limited angle. For larger angles the Hooke's joint was always used, but this again was less satisfactory when the angle between drive and driven shafts became large. This problem arises from the cyclical speed variation that occurs in the driven shaft.

When we consider the case of a drive shaft rotating at constant angular velocity, the driven shaft undergoes a cyclical speed variation every 180 degrees of rotation, being slower than the drive shaft at first then faster during the final 90 degrees of rotation. The mean angular velocity over the $180°$ cycle is, of course, identical to that of the drive shaft. For an angular shift of $8°$ between the lines of the two shafts, the maximum variation in angular velocity is only 2%. If we treble the angle between the shafts to $24°$, the maximum velocity variation increases to 18%. The standard technique of coping with this problem is to use two Hooke's joints, $90°$

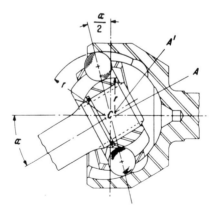

Fig. 11.2 The Rzeppa constant-velocity universal joint.

out-of-phase, as shown in Fig. 11.1. The intermediate shaft still undergoes cyclical variations, these can produce undesirable vibrations that can be transmitted to the body if shaft angles of 20° are exceeded. With FWD much greater angles than 20° are required, if acceptable turning circles are to be given.

The need for a better U/J was apparent for more than a generation and three engineers, Grégoire, Rzeppa and Weiss, all working along similar lines eventually developed *constant-velocity* universal joints that were reliable and compact. A typical Rzeppa joint is shown in cross section in Fig. 11.2. Six hardened steel balls, rolling in grooved raceways in the two halves of the joint, transmit torque across the coupling. The six balls are located by a cage and the purpose of the cage is to maintain the six balls in the same plane, this plane being at an angle, the half angle $\alpha/2$, where α is the relative angle between drive and driven shafts. As the angle α changes, so does the operating plane of the six balls. By maintaining the driving members (the balls) in the median plane, torque remains constant and the drive is therefore at constant velocity. There are variants on this basic design, usually in the profile of the ball tracks. Peugeot now use what is called a 'tripod joint', in which three hardened rollers transmit the drive. The principle is identical to that used by Rzeppa and it also gives constant velocity. Some designs have been developed to accept 'plunge'. In these joints provision is made for lateral movement in a sliding sleeve, since the suspension geometry sometimes calls for a small change in drive shaft effective length.

11.3 The Fiesta suspension

Today, very few cars are designed by one man. Even cars that are made in half-dozen batches like Formula 1 racing cars have a specialist like Keith Duckworth to design the engine. Again, Hewland Engineering Ltd design the transaxle and an overall chassis specialist like Derek Gardner designs the rest. When a mass-produced car such as the Ford Fiesta is designed, relatively large specialist teams are involved with a project leader to co-ordinate every part of the project with the whole. Inevitably, committee efforts are slow but the vast investment involved encourages caution. The executive engineer in charge of chassis design on the Fiesta project was Tony Palmer and his original plan for a wishbone suspension system at the front using torsion bars was changed to the MacPherson strut system, a design used on several earlier British Ford cars.

11.3.1   The front suspension

The general arrangement of the front suspension is shown in Fig. 11.3; while a cross section of the wheel bearing assembly is given in Fig. 11.4. Two aims were dominant in the plans for the Fiesta: ease of assembly and ease of maintenance. The first would benefit both supplier and customer; the second would be very welcome to the customer. A good example of this philosophy is seen in the front wheel bearings. The Fiesta uses Timkin 'Setright' tapered roller bearings. These require no adjustment during assembly, nor the use of an

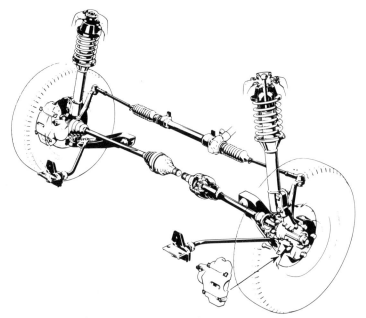

Fig. 11.3   The Fiesta front suspension.

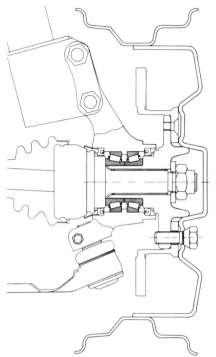

Fig. 11.4 Fiesta front wheel bearing assembly.

assortment of shims of differing thicknesses. These bearings
are installed so that the inner races abut one against the
other. They are not located by the inner diameter. Tight
control on machining tolerances of the hub carrier permits
bearings to be selected at random and to give a preload on
the bearings within acceptable limits when the hub carrier is
torqued on its stub axle shaft. As in most modern designs,
the bearings are prepacked with grease and 'sealed for life'.
This design, then, does appear to meet the two main criteria.
If the bearings should ever need replacement, the elimination
of shims or other preload adjustment should make it a simple
snagfree operation.

The design of the MacPherson strut differs from standard
Ford practice only in the provision of a top bearing. Normally
on British Ford cars rotation of the strut is accommodated by

rotation of the spring and by deflection of the rubber mount that supports the rear spring carrier. To reduce self-aligning torque, Palmer provided a phenolic resin moulding to act as a thrust bearing. To minimize friction, the thrust surface is impregnated with PTFE. Tie rods are provided to give fore-and-aft location of the bottom ends of the struts and to resist brake torque. Rubber bushes are provided at the front of the tie rods and in the transverse arms to give a moderate amount of compliance.

The steering geometry is designed to give a 8-mm 'negative offset'. This is an essential safety feature with a dual braking system designed to provide a diagonal split. With a positive offset, if one brake circuit fails and the car were braked on one front wheel and the diagonally opposite rear wheel, there would be a tendency for the car to steer towards the braked front wheel. A small negative offset counteracts this tendency.

11.3.2   The rear suspension

Effectively the rear suspension is a dead axle located by three main links and two torque straps. The tubular axle is cranked upwards in the middle (see Fig. 11.5) and is located laterally by a Panhard rod. Fore-and-aft location is provided by two channel-section trailing links attached to brackets below the axle tube. A degree of compliance is given by rubber bushes at the forward pivots. A full five-link location would require the provision of two additional longitudinal links, leading or trailing, attached to brackets above the axle tube. Such a system would be required on a modern *live rear axle* with coil spring suspension. A live rear axle, however, must be provided with linkages to resist torque reaction in both directions, since there will be a tendency to rotate the axle tube in one direction under drive torque and in the other direction under braking torque. The Fiesta rear axle will only tend to twist under braking torque and this has been resisted in Palmer's design very simply by the provision of short straps welded to the lower end of each damper tube and extended forwards above the axle tube. Rubber bushes are used to give some compliance where these straps are attached to vertical pins welded to brackets forward of the axle tube.

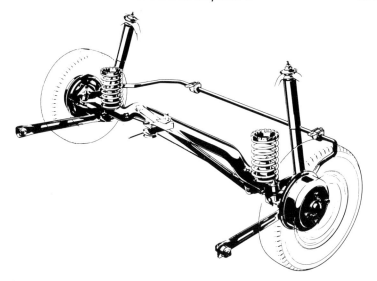

Fig. 11.5 The Fiesta S rear suspension.

Under braking forces axle windup is thus transferred through the straps to the dampers. Piston rods of a diameter larger than would normally be used have been provided to resist this side load. An anti-roll bar is fitted at the rear, but only on the S model.

Suspension data: Ford Fiesta S

Wheelbase,	2.286 m
Front track,	1.334 m
Rear track,	1.321 m
Kerb weight,	727 kg
Laden weight (4-up),	1040 kg
Sprung mass (4-up) (estimated),	945 kg
Front/rear weight distribution (4-up),	54/46
Tyre size,	145-12
Front spring rate, at wheel,	21.7 kN/m

Rear spring rate,* at wheel,	25.0 kN/m
Rear roll bar rate, at wheel,	3.3 kN/m
Total wheel travel, front,	143 mm
Total wheel travel, rear,	166 mm
Wheel travel in bump, front,	62 mm
Wheel travel in bump, rear,	84 mm
Height of roll centre above ground, front,	185 mm
Height of roll centre above ground, rear,	191 mm
Roll stiffness, front,	296 Nm/deg
Roll stiffness, rear,	210 Nm/deg
Roll angle at a lateral acceleration of 0.5g,	3° 54'
Camber angle, static, at kerb weight,	1° 57' positive
Camber angle, static, laden,	0° 52' positive
Toe setting,	2.5 mm, toe-out
Kingpin inclination,	15° 6'.

Height of sprung mass centroid. Since we know the combined roll stiffness at front and rear, and we know the roll angle produced by a lateral acceleration of 0.5 gravity we can calculate the height of the centroid:

Total roll stiffness = 296 + 210 = 506 Nm/deg.
Roll resistance at 3° 54' (3.9°) = 3.9 × 506 = 1973 Nm.
The roll moment (see Fig. 11.6) = $F \times h_1$ = 0.5g $\times M \times h_1$
$$= 0.5 \times 9.807 \times 945 \times h_1$$
$$= 4634 h_1.$$

Therefore

$$h_1 = \frac{1973}{4634}$$
$$= 0.426 \text{ m.}$$

Mean roll centre height = $\dfrac{185 + 191}{2}$ = 188 mm.

Therefore centroid height above ground = 426 + 188 = 614 mm.

---

* This rate is based on both rear wheels in bump and rebound. Single wheel rate is 12% higher.

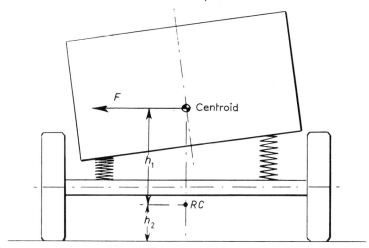

Fig. 11.6

11.3.3 Suspension balance

The following data from Chapter 3 will help us to analyse the overall balance of the Fiesta S suspension in bounce and pitch:

	One-up	Four-up
Front spring frequency, in bounce,	1.55 Hz	1.45 Hz
Rear spring frequency, in bounce,	2.17 Hz	1.69 Hz
Pitch frequency,	1.96 Hz	1.56 Hz
Front wheel distance from centroid l_1	0.846 m	1.052 m
Rear wheel distance from centroid l_2	1.440 m	1.234 m
Front conjugate point distance behind centroid s	1.440 m	1.074 m
Rear conjugate point distance in front of centroid r	0.846 m	1.269 m

The dimensions r, s, l_1 and l_2 are shown in Fig. 11.7. The significance of the double conjugate points r and s was discussed in Chapter 3. On the Fiesta with only the driver in the car the double conjugate points coincide exactly with the wheel centres. This is obviously the designer's choice and means that for this particular loading each 'axle' has no effect

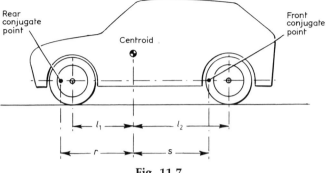

Fig. 11.7

in bounce or rebound on the other 'axle'. Analysis of several modern cars shows this to be a popular choice. As the load is increased, the rear conjugate points moves forward slightly and the front conjugate point moves ahead of the rear axle. Thus any bounce movement at the rear will produce a small amount of bounce at the front *in the same direction*. Since the front conjugate point with four-up falls inside the wheelbase, any bounce at the front will produce a small amount of bounce at the rear, this time *in the opposite direction*. The change from the one-up condition is not great. A bump movement of 40 mm at the front would only result in a rebound movement at the rear of 3 mm. The most significant change from the one-up load to the four-up load is the marked reduction in the rear spring frequency and in the pitch frequency

11.3.4   Roll balance

It has already been stressed that a small family car is subject to a fairly wide variation in load distribution. It is, therefore, difficult to balance the roll resistance under all loads. The roll resistance on the Fiesta S is 296 Nm/deg at the front and 159 Nm/deg from the coil springs at the rear plus 51 Nm/deg from the anti-roll bar. Expressed as a percentage, the total roll resistance is split 58.5% at the front and 41.5% at the rear. This is a very good compromise between the weight distributions of one-up (63/37) and four-up (54/46).

The reader may wonder how the two front springs of 21.7 kN/m rate can exert a greater roll resistance than two rear springs of 25.0 kN/m rate assisted by an anti-roll bar with an effective rate at the wheel of 3.3 kN/m. The answer lies in the effectiveness of the springs in roll, the beam axle being less effective. The roll resistance of a beam axle is proportional to the square of the spring base, whereas the roll resistance of an independently sprung vehicle is proportional to the square of the wheelbase. The rear spring base (the distance between the coil spring centres) is only 72% of the wheelbase. Compared with the front suspension, the rear springs are therefore little more than half as effective in controlling roll.

11.4 The FWD cornering syndrome

In Chapter 6 we illustrated by vector diagrams how an FWD car can, by accelerating through a curve, corner at a higher speed than an RWD car. The danger that could exist when cornering at the limit if the designer has not designed a safety factor into the suspension and steering geometry has also been pointed out. Without careful design, the driver could find himself at the point of no return, where the need to lift his right foot will inevitably result in a spin. One factor on the little Fiesta reduces this danger to some extent: there is very little torque available to spare to exploit the acceleration technique when cornering at speed. If we refer back to Fig. 6.1(b) and (c), we see that the driver of an FWD car can use some of the available torque from the driven wheels to direct the car inwards against the centrifugal force, a technique not available to the driver of an RWD car.

Ever since General Motors were 'persuaded' by Ralph Nader to withdraw the original Corvair from production on the grounds of unsafe behaviour when cornering *when driven by a driver of average ability*, every senior engineer responsible for the production of a new car has been made aware of his ultimate responsibility. The president of the company may say to the Press 'the buck stops here', but the senior design engineer in charge of a new project knows whose job is really at stake! Ironically, a few modifications to the Corvair

196 A small FWD saloon car: Ford Fiesta S

suspension geometry had already made the car perfectly safe, but withdrawal of the model was advisable after so much adverse publicity.

The problem of *power liftoff* on the FWD car is a different one, but it cannot be ignored even on a low-powered car. How then can a suspension engineer build a failsafe feature into the cornering behaviour of an FWD car? When cornering at the limit, a need to slow down for an obstruction ahead could result in a breakaway at the front end. What is needed is a geometry that gives understeer in a corner, but a reduction in understeer (reduced slip angles at the front or increased slip angles at the rear) when the car is decelerated in a bend. There are several ploys available, and ingenious engineers (the adjective and the noun are from the same Latin root) prove their worth every few months by inventing new ones. One recent example from the Porsche engineers is the Weissach axle, described in Chapter 12. This is a suspension design for an RWD car which changes from neutral steer to understeer under liftoff.

Two factors in the steering and suspension geometry of the Fiesta tend to give understeer when under power in a bend and oversteer when the foot is lifted from the accelerator, or even under a moderate brake application. The first factor is the position of the steering rack. If the steering rack is mounted behind the front wheels, lateral forces cause the bushes to deflect in such a way that a small amount of oversteer is produced. This is the position of the steering rack on the Fiesta. Roll understeer can be designed into a trailing link located rear axle by the inclination of the links as shown (much exaggerated) in Fig. 6.6. With the forward pivots of the trailing links below wheel hub height, the outer wheel moves forward when rolling in a bend and the inner wheel moves backward: this gives understeer.

Let us consider then what happens when acceleration in a fast bend is suddenly followed by deceleration. For stable cornering, the balance between front and rear tyre forces and roll moments have been carefully balanced. Front slip angles will be a little greater than those at the rear. As soon as the brakes are applied (even a driver of average ability should know not to stamp on them in the middle of a bend), there is a transfer of load from the rear to the front. This increases the lateral forces on the steering bushes, and with sufficient compliance this oversteering effect reduces the overall understeer. The second effect of braking is that the reduction in load on the rear axle gives a reduction in the rear roll angle. This again gives a reduction in understeer. It is doubtful if the combined effect of these two factors would make any FWD car completely 'clotproof' on a slippery bend, but the experience of the motoring Press, with many miles of testing in the severe British winter of 1978–9, has been unanimous. The handling of the Fiesta is safe and predictable.

12

A High-performance Sports Car: Porsche 928

12.1 Porsche panache

The Porsche Co. has a long history of development in the field of suspension and handling, some of this won the hard way by attempting to solve problems introduced by the inherent weaknesses in their initial design.*

Many readers will be familiar with the Porsche 911, that brilliant descendant of a long line of sports cars with air-cooled engines at the rear. The evolution of semi-trailing link suspensions at both front and rear has been finely tuned to give excellent handling.

The new models emanating from the Porsche Design Office, the 2-litre 924 and the 4.5 litre 928, are most surprisingly front-engined cars with rear wheel drive. The final shock is the switch from air-cooled to water-cooled engines, but this is a subject outside the scope of this book.

Helmut Bott, the Director of Engineering at Porsche, has, stated that the change to a front-engined location was made to ease the problems associated with the USA 30 m.p.h. simulated front-end collision test. By using a rigid tube to connect the engine and rear-mounted transaxle, the impact energy can be absorbed by the main mechanical components. Fig. 12.1 is a ghosted view of the Porsche 928. The engine central tube and transaxle can be seen in the photographic plan in Fig. 12.2.

* Dr Porsche fought doggedly with the inherent handling problems of swing axle suspension. On the 520 b.h.p. Auto Union P-Wagen GP car the problems were almost insuperable.

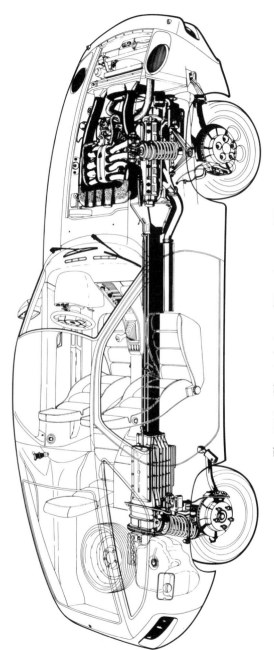

Fig. 12.1 Ghosted view of Porsche type 928.

Fig. 12.2   Porsche 928 with body removed.

12.2   The Porsche 928 suspension

Upper and lower wishbones are used at the front. At the rear upper and lower transverse links are used with the addition of a special patented flexibly mounted trailing link known as the Weissach-Axle. The suspension geometry gives 30% anti-dive at the front and 50% at the rear. The rear suspension layout also gives 70% anti-squat. Compromise is essential in these factors, especially when designing a sports car, if adverse changes in toe-in, camber and castor angles are to be avoided. The suspension engineers at Weissach, the Porsche research centre, admit that they arrived at the chosen geometry by trial and error. With a new untried design one can hardly expect the computer to give a reliable answer.

On the front suspension an alloy casting is used to provide a conventional upper wishbone. An even more robust alloy casting is used for the transverse lower arm. These components are particularly robust since they are subjected to very high forces during braking, acceleration and cornering. By comparison the suspension components on the Fiesta are quite flimsy. The Porsche 928, however, has a power-to-weight ratio of 115 kW/tonne (160 b.h.p./ton) and the Pirelli P7 225/50 VR16 tyres, fitted as standard, endow the car with phenomenal cornering accelerations. The use of tyres with such a low-profile ratio does, however, involve the penalty of a suspension geometry that allows only small changes in wheel camber. At the front the variation from full bump to full rebound is only about 1.5°. At the rear it is less satisfactory since it approaches 4 degrees.

The components of the rear suspension can be seen in Fig. 12.3, which is a photograph of the rear suspension cross-member and the suspension components taken from the front. Each wheel hub carrier is located by a simple transverse upper link and a more complex lower link which is, in effect, an integrated two-part link. The transverse member is a relatively thin steel plate, very stiff under pure vertical loads, but flexible in the fore-and-aft direction. The trailing link is attached to an articulated double-pivot, which lies approximately in line with the inner pivot (the pivot of the steel plate) to make an angle of approximately 25° with the centre line of the car. The Weissach-Axle is, therefore, a semi-trailing link suspension with a variable axis. The dangers of excessive amounts of compliance in the rubber bushes used in suspension links are well known to suspension engineers. In racing car design, where small variations in suspension geometry cannot be tolerated, rubber bushes are never used. Drivers of ordinary passenger vehicles do not expect to suffer shocks, vibrations or noise transmitted directly from the suspension attachments. Isolation, within

Fig. 12.3 Porsche rear suspension, viewed from the front.

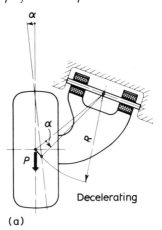

(a)

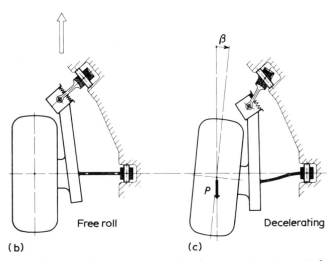

(b)   (c)

Fig. 12.4 (a), Porsche 928 conventional suspension toe-out due to elasticity. When decelerating or braking, the wheels in a conventional suspension toe-out at α angle owing to force P. The reasons are the rubber bearings in the suspension required for noise absorption. (b)–(c), Weissach-Axle – toe-in due to kinematics. A kinematic effect changes the angle towards toe-in when decelerating or braking. For reasons of noise absorption, rubber bearings are featured, but toe-out is immediately balanced by the kinematic effect $\alpha = \beta$. Thus dangerous side movements are prevented in curves, an important characteristic.

the limits of current technology, is therefore the order of the day, even on a sports car.

Cunning old Americans say 'If you can't lick 'em, join 'em'. This is the philosophy of the Weissach-Axle. Rubber mountings are used on the Porsche 928 to modify suspension behaviour under traction, free roll and deceleration to give an automatic correction to the amount of rear wheel toe-in. Even when cornering near the limit, the fine tuning of the variation in toe-in calls for negligible steering correction from the driver when forced to decelerate.

With greatly exaggerated angles, the Weissach-Axle principle is explained in Fig. 12.4. On a conventional semi-trailing link rear suspension deceleration (and to a greater extent, braking) deflects the rubber-mounted pivots to give a toe-out angle to both rear wheels. This is an oversteering effect. Since braking also robs the rear tyres of effective cornering force and increases the rear slip angles (see the Circle of Forces, Fig. 1.17), these two oversteering effects in combination can sometimes give the driver a difficult handling problem when decelerating in a corner. To the professional test driver, this is often termed *liftoff tuck-in*. When very rapid unwinding of the steering wheel is necessary to correct this oversteer, the car can hardly be called 'inherently stable'.

The front pivot of the Weissach-Axle is shown schematically in Fig. 12.4. It consists of two bushes, a conventional rubber bush on a rod end and a larger rubber bush, 100 mm long and 50 mm diameter, to which the rod end is fixed. Only very small changes in the amount of toe-in are permitted by this double-jointed pivot system. The static toe-in is 20', decreasing under traction, but never reaching a toe-out angle. Under severe braking, the toe-in will increase but will seldom reach more than 50'. These are very small angular changes, but sufficient to give vicefree cornering behaviour.

The following data have been supplied by Porsche:

Type 928 chassis

Wheelbase,	2.5 m
Track, front,	1.545 m
Track, rear,	1.514 m

Kerb weight,	1450 kg
Permissible total weight,	1870 kg
Weight distribution, one-up,	49/51
Weight distribution, two-up plus luggage,	47/53

Type 928 suspension

	Front	Rear
Wheel rate,	18.63 kN/m	22.55 kN/m
Anti-roll bar rate, at wheel,	83.4 kN/m	10.2 kN/m
Total roll stiffness,	2115 Nm/deg	655 Nm/deg
Wheel travel, bump,	90 mm	100 mm
Wheel travel, rebound,	90 mm	100 mm
Roll centre height above ground,	70 mm	110 mm
Camber angles		
full bump,	$-2°\ 8'$	$-4°\ 30'$
normal laden,	$-30'$	$-1°$
full rebound,	$-42'$	$-33'$
Castor angle,	$3°\ 30'$	
Kingpin inclination,	$13°\ 36'$	
Toe-in, static,	0	$20'$

12.2.1   Yaw response

Modern wide-section tyres have exerted a profound influence on the thinking of designers of sports cars. Twenty years ago one could argue with no thought of contradiction, that a sports car should have a low polar moment of inertia, this being the polar moment about the vertical axis as discussed in Chapter 6. At that time it was contended that a rapid change of direction, such as a lane-change on a motorway or a chicane on a race circuit, could be more accurately and safely executed when the polar moment of inertia was low. Today, the grip of modern tyres with their ultra-low profiles and much-improved rubber compounds makes it difficult to establish which sports cars have the least tendency to spin, since very rapid changes of direction can be made even in the wet by sports cars with both high and low polar moments of inertia. Experiments were carried out in 1977 (*Motor*, May

7th 1977) on a range of sports cars with polar moments of inertia (measured at the Cranfield Institute of Technology) varying between 1112 kg/m^2 for the little mid-engined Fiat X1/9 to 2050 kg/m^2 for the Porsche Carrera. In these experiments handbrakes were applied to lock the rear wheels then released immediately, while the car was in the middle of a 50-m radius turn. The wetted road surface had a coefficient of friction of approximately 0.3. The curve was negotiated at a speed of 30 m.p.h. This technique (reminiscent of the handbrake tactics used by rally drivers when they require to put their car sideways-on as they approach a tight turn) inevitably made the rear end of every sports car tested breakaway and begin to spin. The cars were all equipped with instruments to measure the *yaw acceleration*, since it was reasonable to suppose that a car that spun with the highest rate of angular acceleration would be the one that would be the most difficult to control by steering correction.

The experiments were inconclusive. No correlation was apparent between maximum yaw accelerations and polar moments of inertia. The 'seat of the pants' impressions of the drivers gave no indication that the tendency to spin on a slippery surface was any greater with front-engined or rear-engined cars or with those of low or high polar moment of inertia. It is not surprising, therefore, that the Porsche design team working on the type 928 had no hesitation in adopting a completely different layout of the major masses than they had used in the past. Since they planned the chassis on the basis of identical low-profile tyres at front and rear, they saw that the essential requirement was a mass distribution that gave almost equal static tyre loads. This has been achieved with a 49/51 weight distribution with a one-up load and a 47/53 distribution with two-up plus luggage.

With the phenomenal grip offered by the Pirelli P7 tyres the second essential requirement was a close control on camber angle change. This is achieved on the front suspension with a variation between full bump and full rebound of only 1.5 degrees. The camber variation at the rear is greater, varying between −4 degrees 30 minutes at full bump and −33 minutes at full rebound. This represents the *maximum* change of camber between bump and rebound stops and is no measure of the camber changes experienced in roll.

12.2.2 Roll

Roll is controlled on the Porsche 928 by the provision of stiff anti-roll bars, in particular at the front. Since the roll centre is 40 mm lower at the front than at the rear (see Suspension Data, above), one would anticipate the need for a higher roll resistance at the front. The roll resistance at the front is actually about three times that at the rear.

The combined roll stiffness of suspension springs plus both anti-roll bars is 2770 Nm/deg. We have taken an estimated value of the sprung mass centroid height at a value of 440 mm. Using this, we can make an estimated approach to the roll angle under a chosen lateral acceleration.

Using the same nomenclature as in Chapter 11 (see Fig. 11.6):

$$h_2 \text{ (mean)} = 90 \text{ mm}$$
$$h_1 \text{ (estimated)} = 440 - 90 = 350 \text{ mm}$$

The roll moment $= F \times h_1$.

At 1 gravity centrifugal acceleration, the roll moment

$$RM = 1.0 \times 9.807 \times 1385 \times 0.350 = 4754 \text{ Nm}$$

where 1385 kg is the estimated sprung mass (one-up). Since the roll stiffness is 2770 Nm/deg, the mean roll angle at 1 gravity cornering acceleration is:

$$\frac{4754}{2770} = 1.7° = 1° \, 42'.$$

It is interesting to compare this with the Ford Fiesta S in Chapter 10 which rolls through $3°54'$ under 0.5 gravity centrifugal acceleration.

12.2.3 Camber change in roll

A roll of $1°42'$ on the type 928 will give a bump movement on the outer rear wheel of $\sin(1°42') \times 757$ mm, where 757 mm is half the rear track.

This is a bump movement of 22.7 mm (about one-quarter of the available bump travel). This will increase the normal laden camber angle at the rear from $-1°$ to about $-1°48'$. The camber changes on the inner rear wheel and the two front wheels will be negligible. Thus by the provision of good roll resistance, Porsche have met this essential requirement for a car fitted with 50 series tyres.

12.3   Natural frequencies and double conjugate points

Using Professor Guest's method as outlined in Chapter 3, we have sufficient data to make a close approximation to the natural suspension frequencies and double conjugate points with different laded weights:

Frequency data (one-up)

Total mass M_t	= 1525 kg
Sprung mass M_s (estimated)	= 1385 kg
Wheelbase	= 2.5 m
Weight distribution	= 49/51
l_1 = 0.51 × 2.5	= 1.275 m
l_2 = 0.49 × 2.5	= 1.225 m
k_1	= 18.63 kN/m
k_2	= 22.55 kN/m
$a = \dfrac{k_2}{k_1 + k_2} \times 2.5$	= 1.369 m
$b = \dfrac{k_1}{k_1 + k_2} \times 2.5$	= 1.131 m
$\dfrac{K^2}{l_1 \times l_2}$ (estimated*)	= 0.98

Therefore

K^2	= 1.5306
$c^2 = a \times b$	= 1.548
$x = a - l_1$	= 0.094 m
x^2	= 0.0088.

From this data we can obtain the following values:

Conjugate Points (one-up)
Front conjugate point distance behind centroid s = 1.382 m
Rear conjugate point distance in front of centroid r = 1.103 m.

---

*   For the type 924 (the 2-litre model) Cranfield Institute measured a value of 0.95.

Natural frequencies (one-up)

F_f = 1.18 Hz
F_r = 1.28 Hz
F_{pitch} = 1.24 Hz.

Frequency data (two-up plus luggage)

Total mass M_t	= 1800 kg
Sprung mass M_s (estimated)	= 1660 kg
Wheelbase	= 2.5 m
Weight distribution	= 47/53
l_1 = 0.53 × 2.5	= 1.325 m
l_2 = 0.47 × 2.5	= 1.175 m
k_1	= 18.63 kN/m
k_2	= 22.55 kN/m
K^2	= 1.5306
a	= 1.369 m
b	= 1.131 m
c^2	= 1.548
x = $a - l_1$	= 0.044 m
x^2	= 0.001936.

This data for the more heavily loaded condition gave the following values:

Conjugate points (two-up plus luggage)

s = 1.476 m
r = 1.037 m.

Natural frequencies (two-up plus luggage)

F_f = 1.10 Hz
F_r = 1.14 Hz
F_p = 1.13 Hz.

12.4 Comments

For a sports car the natural frequencies of the suspension, both front and rear, are exceptionally low. Over the majority of road surfaces the ride cannot fail to be good, with much lower vertical accelerations than one finds in many modern sports cars. Since Porsche have chosen to make front and rear natural frequencies almost identical, it is inevitable that the natural frequency in pitch is also very close to these values. With much higher frequencies – as, for example, those used for the Ford Fiesta – it would be inadvisable to permit such a close approach, but the use of a well-damped suspension in the Porsche has reduced the dangers of pitch and bounce movements overlapping to give any discordant 'joggle' to the ride.

One penalty associated with the use of such a stiff anti-roll bar as that used at the front of the Porsche 928, is that the spring rate for single wheel bump is much higher than that for double wheel bump. The frequency in single wheel bump is controlled by three springs. Of course, the first spring is the main spring on the side involved. Added to this is the spring rate of the anti-roll bar *acting in series* with the main spring on the other side of the car.

As discussed in Chapter 4, two springs working in series behave as if replaced by an equivalent spring S_e, where

$$S_e = \frac{S_1 \times S_2}{S_1 + S_2} \, .$$

The normal wheel rate at the front is 18.63 kN/m and the anti-roll bar rate is 83.4 kN/m. The additional equivalent spring rate is therefore:

$$S_e = \frac{83.4 \times 18.63}{83.4 + 18.63} = 15.23 \text{ kN/m}.$$

The total spring rate for single wheel bump at the front is therefore 18.63 + 15.23 = 33.86 kN/m.

Since the natural frequency of a spring system varies as the square root of the effective spring rate, the natural frequency for single wheel bump at the front is 35% higher than the frequency for double wheel bump. The rear anti-roll bar is not very stiff and the single wheel bump frequency at the rear is only about 14% higher than the double wheel bump frequency.

The noticeable hardening of the front suspension under single wheel bump has received some adverse comments from a few members of the motoring press. On the whole, it appears to be a small price to pay for a car that rides so well and corners like a real sports car – and that undoubtedly is the best description of the Porsche 928.

Index

02311